COSMOS & TRANSCENDENCE

*Breaking Through
the Barrier of Scientistic Belief*

Wolfgang Smith

COSMOS &
TRANSCENDENCE

*Breaking Through
the Barrier of Scientistic Belief*

SECOND EDITION

SOPHIA PERENNIS

SAN RAFAEL, CA

Second, revised edition, 2008
First edition,
Sherwood Sugden & Co., 1984
© Wolfgang Smith 2008
All rights reserved

For information, address:
Sophia Perennis, P.O. Box 151011
San Rafael CA 94915
sophiaperennis.com

Library of Congress Cataloging-in-Publication Data

Smith, Wolfgang
Cosmos and transcendence: breaking through the
barrier of scientistic belief / by Wolfgang Smith.—2nd ed.

p. cm.

Rev. ed. of: Cosmos & Transcendence. c1984
Includes bibliographical references and index.
ISBN 978-1-59731-080-2 (pbk: alk. paper)
ISBN 978-1-59731-084-0 (cloth: alk. paper)
1. Science—philosophy. I. Smith, Wolfgang,
Cosmos & transcendence. II. Title
Q175.S636 2008
501—dc21   2008005657

ACKNOWLEDGMENTS

The author and publisher gratefully acknowledge permission to reproduce the frontispiece of Codex 2554, a thirteenth-century Bible housed in the Austrian National Library in Vienna.

Grateful acknowledgment is also made for permission to reprint excerpts from the following material:

Douglas Dewar, *The Transformist Illusion*, © 1995 by Sophia Perennis.

Louis Bounoure, *Déterminisme et finalité*, © 1957 by Flammarion. Reprinted with permission of Librairie Ernest Flammarion, Paris.

C. G. Jung, *Modern Man in Search of a Soul*, © 1933 by Harcourt Brace Jovanovich, Inc. Reprinted with permission of Harcourt Brace Jovanovich, Inc., New York.

Kant's Critique of Pure Reason, trans. by Norman Kemp Smith, © 1958 by Random House, Inc. Reprinted with permission of St. Martin's Press, New York.

Sigmund Freud, *An Outline of Psychoanalysis*, © 1949 by W. W. Norton & Company. Reprinted with permission of W. W. Norton & Company, New York.

Sigmund Freud, *The Ego and the Id*, © 1960 by W. W. Norton & Company. Reprinted with permission of W. W. Norton & Company, New York.

Sigmund Freud, *New Introductory Lectures on Psychoanalysis*, © 1965 by W. W. Norton & Company. Reprinted with permission of W. W. Norton & Company, New York.

C. G. Jung, *Memories, Dreams, Reflections*, recorded and edited by Aniela Jaffe, trans. by Richard and Clara Winston, © 1963 by Pantheon Books, a Division of Random House, Inc. Reprinted with permission of Random House, Inc., New York.

*For Thea, whose sound judgment
has so many times saved the day*

CONTENTS

FOREWORD

As THIS BOOK makes compellingly clear, the Scientific Revolution of the seventeenth century heralded the triumph of a particular philosophical outlook (rationalistic, materialist) with its attendant epistemology (empiricism) and procedures (the 'scientific method'). Contrary to popular assumption, modern science is *not* simply a disinterested, detached, and value-free mode of inquiry into the material world: it is a complex of disciplines and techniques anchored in culture-bound *assumptions* about and *attitudes* to the nature of reality and the proper means whereby material phenomena might be explored, explained, and, perhaps most tellingly, controlled. It is, in fact, impossible to separate the methods of modern science from its theories and the ideologies which provide its motive force, and it is to this tangled skein—or as he puts it: to the unverifiable assumptions which the 'verifiable' propositions of science presuppose—that Wolfgang Smith applies the term 'scientism'.

As the book likewise makes clear, the modern scientistic outlook is, *in principle*, unable to countenance God (by whatever name), who is either repudiated as an obsolete 'hypothesis' or altogether ignored—which amounts to the same thing. Likewise, scientism cannot allow any *sense of the sacred*, the absence of which is one of the defining characteristics of modernity as a whole. Needless to say, the much-misunderstood issues at stake in the conflict of 'science' and 'religion', or 'modernity' and 'tradition', are immense: our view of what constitutes 'reality', 'human nature', 'life' and 'death', transcendence and immanence, and the relationship of the material world to higher spiritual realities, to mention only some of the most salient. It is to an inquiry into these issues—an interrogation of the orthodoxies of modern science in the light of a traditional wisdom, informed by immutable principles and truths which are neither 'old' nor 'new' but timeless—that *Cosmos and Transcendence* summons us.

Wolfgang Smith brings to his task a rare combination of qualities and experiences, not the least his ability to move freely between the somewhat arcane worlds of contemporary science and traditional metaphysics. Alongside Dr. Smith's imposing qualifications in mathematics, physics, and philosophy, we find his hard-earned expertise in Platonism, Christian theology, traditional cosmologies, and Oriental metaphysics. His outlook has been enriched both by his diverse professional experiences in the high-tech world of the aerospace industry and in academia, and by his own researches in the course of his far-reaching intellectual and spiritual journeying. Here is that rare person who is equally at home with Eckhart and Einstein, Heraclitus and Heisenberg! Dr. Smith is no obscurantist, rejecting well-attested scientific facts, nor a sentimental reactionary seeking to 'turn back the clock'. He is a sober-minded scientist and philosopher who has confronted some of the most daunting issues of the age, refusing to surrender to the shibboleths and complacencies of modernity.

In this book Wolfgang Smith excavates the very foundations of modern thought in order to explain the cracks and fissures that are everywhere appearing in what was thought to be the impregnable edifice of 'science'. He also traces the pedigree of some of the most mesmerizing of modern prejudices (the belief in Progress, for instance) and analyzes the intellectual legacy of figures such as Descartes, Newton, Darwin, Freud, and Jung, all the while rendering the most abstruse ideas and principles into lucid and elegant prose, intelligible to any receptive reader. *Cosmos and Transcendence*, which first appeared a quarter of a century ago, is the fruit of many years of fearless intellectual exploration, of deep rumination, and of seasoned judgment. Our era stands in urgent need of the truths and insights yielded by Wolfgang Smith's wide-ranging inquiry. Sophia Perennis is to be commended for bringing a new edition of this profound and exhilarating work within the purview of a new generation of readers.

HARRY OLDMEADOW
La Trobe University
Bendigo, Australia

PREFACE

TO SECOND EDITION

THIS BOOK HAS A TWOFOLD PURPOSE, a twofold content: first, it presents a critique of the modern world, and based upon this critique, it seeks to expound a timeless metaphysical wisdom. The second end presupposes the first: so long as we have not 'broken through the barrier of scientistic belief,' as the subtitle has it, that timeless wisdom—that veritable *sophia perennis*—remains for us inaccessible.

What I object to fundamentally in the scientistic world-view is that it conceives the external universe to be unperceived and unperceivable. The concrete world, made up of sensory elements, such as color and sound, and indeed of innumerable qualities, is thereby subjectivized, which is to say that it is relegated to the sphere of mind, or if you will, of brain-function. Now, in keeping with major philosophic trends (beginning with Husserl and Whitehead), I regard this subjectivization as both illegitimate and grossly deceptive. My concern, in *Cosmos and Transcendence*, has been to show, on the one hand, that the subjectivization of the qualities is not a matter of scientific discovery, as one is nowadays prone to assume, but constitutes in fact an unfounded philosophical assumption stipulated by René Descartes, and to demonstrate, on the other, that this Cartesian premise contradicts the perennial wisdom of mankind. On this double basis I could proceed to carry out the twofold intention of the book, as defined above.

And so the matter stood until, a few years later, I became interested in the so-called 'quantum reality' debate, which had been ongoing since about 1927. What has disturbed physicists and philosophers of science for all these years is that the discoveries of quantum theory do not accord with our accustomed notions concerning physical reality, to the point that these findings strike us as paradoxical. What particularly interested me was to ascertain whether the resources of traditional philosophy—I was thinking especially of the Platonist schools—might have something of value to contribute to the debate; and what I discovered, after a period of considerable confusion, came as a surprise: the key to an understanding of quantum theory, I now perceived, lies precisely in the recognition that the qualities are *not* after all subjective, as everyone engaged in the debate had all along assumed. It turns out that once this unfounded Cartesian premise is jettisoned, everything falls into place, and I could write, in *The Quantum Enigma*, that 'Quantum paradox proves to be Nature's way of refuting a spurious philosophy.'

Thus it came about that what, in *Cosmos and Transcendence*, had served as the means of disqualifying the scientistic Weltanschauung, has become key to a philosophic understanding of contemporary physics. The fact is that physics *can be* interpreted on a non-Cartesian basis, and that this reinterpretation constitutes the mandatory rectification which enables us to integrate its positive findings into higher realms of knowing. The very science, thus, which since its inception in the seventeenth century had presented itself as hostile to the traditional wisdom, proves now to be in a way supportive of its claims.

There is however more to be said, for it happens that the aforesaid reinterpretation of physics has a vital bearing on just about every fundamental domain of contemporary science. At the risk of speaking in overly condensed terms, and thus incomprehensibly, I will cite a few examples: (1) The new understanding of quantum theory reveals a principle of 'vertical causation'—a causation which is instantaneous and not determined by antecedent events—which proves to be operative, not only in what physicists term state vector collapse, but in every domain—e.g., that of human art—to which the notion of 'intelligent design' is applicable.[1] (2) The ontological distinction of the *physical* and the *perceivable* environment entails a distinction between the *terrestrial* and the *sidereal* cosmos, which

---

1. See *The Wisdom of Ancient Cosmology* (Oakton, VA: Foundation for Traditional Studies, 2003), chap. x.

fundamentally disqualifies the reductionist claims of contemporary astrophysical cosmology.[2] (3) In a universe endowed with real qualities, what has been termed the anthropic principle assumes an entirely new and unsuspected significance.[3] (4) The deconstruction of the Cartesian premise vitally affects the problem of how we perceive, and supports the empirical findings of James Gibson, the Cornell University scientist who astounded the erudite communities with his so-called 'ecological' theory of visual perception.[4] (5) The aforesaid deconstruction likewise affects the mind-body problem in the context of neurophysiology—the so-called 'binding problem'—and permits an integration of neurophysiological findings into the traditional anthropologies.[5]

Let this much suffice to indicate that the removal of the Cartesian premise, and the resultant return to metaphysical normalcy, proves to be seminal in the extreme. What I wish to convey to the reader in this updated Preface is that the book at hand is to be viewed, not so much as the completion of an inquiry, but as a new beginning, a point of departure in the quest for truth.

CAMARILLO, CA
January, 2008

2. Op. cit., chap. vii.
3. Op. cit., chap. xi.
4. See 'The Enigma of Visual Perception', *Sophia*, vol. 10, no. 1, 2004.
5. 'Neurons and Mind', *Sophia*, vol. 10, no. 2, 2004.

1

THE IDEA OF THE
PHYSICAL UNIVERSE

NOTHING APPEARS to be more certain than our scientific knowledge of the physical universe. But then, what *is* the physical universe? We are told that it consists of space, time, and matter, or of space-time and energy, or perhaps of something else still more abstruse and even less imaginable; but in any case we are told in unequivocal terms what it excludes: as all of us have learnt, the physical universe is said to exclude just about everything which from the ordinary human point of view makes up the world. Thus it excludes the blueness of the sky and the roar of breaking waves, the fragrance of flowers and all the innumerable qualities—half-perceived and half-intuited—that lend color, charm and meaning to our terrestrial and cosmic environment. In fact, it excludes everything that can be imagined or conceived, except in abstract mathematical terms.

But where does this leave our familiar habitat: this ordinary, unsophisticated world, which artists have painted and poets have sung? Are there conceivably two worlds: a visible domain, let us say, and in addition, the physical universe, which science alone can unveil? Obviously we do not misrepresent the quasi-official doctrine if we answer that there is only one real and objectively existing world, which is in fact precisely the physical universe, and nothing more. This (one and only) world, moreover, though it is the cause of perception, is not itself perceived. For what is given in the act of perception (in the sense of an immediate presentation, such as redness, for example) is held to be private and subjective, and so in a way, illusory. Whatever else these 'mental images' may be, they have no place within the physical universe, and in consequence, no real or objective existence. From time immemorial, mankind has apparently been duped by its senses, in that it has attributed to the external world a host of qualities which the world does not possess. As Alfred North Whitehead has put it:

> Thus nature gets credit which should in truth be reserved for ourselves: the rose for its scent: the nightingale for his song: and the sun for his radiance. The poets are entirely mistaken. They should address their lyrics to themselves, and should turn them into odes of self-congratulation on the excellency of the human mind. Nature is a dull affair, soundless, scentless, colorless; merely the hurrying of material, endlessly, meaninglessly.[1]

This is the familiar and yet perennially astonishing hypothesis which stands at the heart of the scientific Weltanschauung: the concept of bifurcation (to use Whitehead's term). More explicitly, what is being bifurcated or cut asunder are the so-called primary and secondary qualities: the things that can be described in mathematical terms, and the things that cannot. Logically speaking, the bifurcation postulate is tantamount to the identification of the so-called physical universe (the world as conceived by the physicist) with the real world *per se*, through the device of relegating all else (all that does not fit this conception) to an ontological limbo, situated outside the world of objectively existent things. The postulate thus eliminates at one stroke precisely those aspects of the world which prove to be recalcitrant to mathematical description: all elements, that is, which cannot be reduced to extension and number. This leaves an inherently mathematical universe, the very thing which a science based upon measurement and calculation could hope to master. In other words, it leaves what we have called the physical universe, not as a mere abstraction or a useful model, but as the objective reality itself. Right or wrong, let it be said at once that this reduction of the world to the categories of physics is not a scientific discovery (as many believe), but a metaphysical assumption that has been built into the theory from the outset.

---

1. *Science and the Modern World* (New York: Macmillan, 1953), p54.

In fact, the thesis traces back to Galileo and Descartes (as we shall have occasion to see in Chapter 2). From thence it was transmitted to Newton, who took over the basic metaphysical conceptions of his continental peers, and incorporated them into the *Principia*, often in the form of scholia appended to his scientific theorems. And from there, of course, it was injected into the mainstream of scientific thought.

But let us not fail to observe that in the course of this transmission something very remarkable has happened to the doctrine. Thus, on the one hand, we find Newton proclaiming the new bifurcationist metaphysics with all the enormous weight of his scientific authority, to the point of engaging himself in elaborate disputations (in the *Opticks*) to demonstrate that the 'secondary qualities' arise within the soul or 'thinking substance,' which he conceives as being situated within a small chamber of the brain (the so-called sensorium); and yet, in numerous other places, 'when his empiricism is not forgotten,' as Edwin A. Burtt has observed, 'Newton speaks of man as being in immediate perceptual and knowing contact with physical things themselves—it is they that we see, smell and touch.'[2] What is still more astonishing, he is willing to extrapolate this sense knowledge even into the atomic realm, as we learn from the following passage in the *Principia*:

> We no other way know the extension of bodies than by our senses, nor do these reach it in all bodies; but because we perceive extension in all bodies that are sensible, therefore we ascribe it universally to all others also. That abundance of bodies are hard, we learn by experience; and because the hardness of the whole arises from the hardness of the parts, we therefore justly infer the hardness of the undivided particles not only of the bodies we feel but of all others. That all bodies are impenetrable, we gather not from reason, but from sensation.[3]

2. *The Metaphysical Foundations of Modern Physical Science* (New York: Humanities Press, 1951), p230.

3. *The Mathematical Principles of Natural Philosophy* (London: 1803), II, 161; cited in Burtt, (n2 above), p229.

Moreover, Newton the empiricist engages in ceaseless polemics against what he calls 'hypotheses', by which he understands any and all affirmations not derived from sensible phenomena and supported by carefully conducted experiments. He perceives his own theories as belonging to 'experimental philosophy', a discipline which he assumes to be totally irreconcilable with 'hypotheses' of any description. The matter is clearly enunciated in the *Principia*:

> Whatever is not deduced from the phenomena is to be called hypothesis; and hypotheses, whether metaphysical or physical, whether of occult qualities or mechanical, have no place in experimental philosophy. In this philosophy particular propositions are inferred from the phenomena, and afterwards rendered general by induction. Thus it was that impenetrability, the mobility, and impulsive force of bodies, and the laws of motion and of gravitation, were discovered.[4]

In short, the Newtonian heritage turns out to be multifaceted and curiously equivocal. Apart from mechanics, optics, and gravitational theorems, it contains the elements of Cartesian metaphysics and an uncompromising positivism, all brought together in one *magnum opus* of incalculable influence. There can be no doubt that the bifurcation tenet has profited greatly from these associations. As Burtt has expressed it, 'Magnificent, irrefutable achievements gave Newton authority over the modern world, which, feeling itself to have become free from metaphysics through Newton the positivist, has become shackled and controlled by a very definite metaphysics through Newton the metaphysician.'[5] To suggest, however briefly, the broader implications of that 'very definite metaphysics' which thus imposed itself upon the modern world, we would like to quote one final passage from Burtt's treatise:

> Wherever was taught as truth the universal formula of gravitation, there was also insinuated as a nimbus of surrounding belief that man is but the puny and local spectator, nay irrele-

---

4. *Principles*, II, 314; cited in Burtt, op. cit., p214.
5. *The Metaphysical Foundations of Modern Physical Science*, p227

vant product of an infinite self-moving engine, which existed eternally before him and will be eternally after him, enshrining the rigor of mathematical relationships while banishing into impotence all ideal imaginations; an engine which consists of raw masses wandering to no purpose in an undiscoverable time and space, and is in general wholly devoid of any qualities that might spell satisfaction for the major interests of human nature, save solely the central aim of the mathematical physicist.[6]

TOWARDS THE END of the nineteenth century, when the victory of Newtonian physics (along with its 'nimbus of surrounding belief') seemed virtually assured, certain unexpected difficulties began to crop up. The remarkable progress of physics, combined with the development of modern technology and the evolution of scientific instruments, had paved the way for certain sensitive experiments, the results of which seemed not to fit the existing theory. Efforts to modify the theory by means of *ad hoc* hypotheses led invariably to less than satisfactory results. Eventually, as we know, Newtonian physics was perforce abandoned as a fundamental or primary theory, even though it survives in a limited capacity (as the appropriate theory for dealing with a certain intermediate or 'mesocosmic' range of physical reality). Curiously enough, it is the very power of that theory—those incredible precisions that had all but converted the world to the Newtonian doctrine—which in the end precipitated its downfall.

The difficulties in question prompted some of the leading physicists to re-examine the foundations of Newtonian physics with great care. Under the influence of logical positivism and kindred philosophical schools, efforts were made to clarify the relation between fundamental physical concepts and observable facts. After centuries of Newtonian domination, it began to dawn upon the bolder spirits that physics does not in fact deal with absolute entities, standing

6. Ibid., p 299.

forever behind the veil of observed and observable Nature, but that—quite to the contrary—physics deals precisely with what is or could be observed through specified physical procedures. After more than two hundred years, physicists took up once more the fight against 'hypotheses', to discover that Newtonian physics was not after all the pure 'experimental philosophy' which it had claimed to be. As Eddington has put it, 'relativity theory made the first serious attempt to insist on dealing with the facts themselves. Previously scientists professed profound respect for the "hard facts of observation"; but it had not occurred to them to ascertain what they were.'[7]

Of course, what is or is not a 'hard fact of observation' depends very much upon the range of magnitudes with which one deals, and upon the sensitivity of one's measuring instruments. In a way, classical physics too was in touch with hard facts, as everyone knows well enough. Its precision (in that sense) was altogether adequate to the domain of applications with which it was primarily concerned. Where classical physics was strangely deficient (and this, surely, is what Eddington has in mind) is in the understanding of its own methods, as evidenced by the ability to provide a clear and coherent account of its actual *modus operandi*. What is more, there was virtually no awareness of this deficiency. Throughout the Newtonian era, a nimbus of confused notions had camouflaged the difficulty, and a mystique of infallibility had in turn upheld that nimbus. To the very end, classical physics perceived itself as a rationally coherent structure, resting squarely upon the bedrock of empirical fact.

As WE KNOW, that perception has changed in consequence of a critical analysis (a kind of scientific epistemology) which began to be pursued in earnest during the early decades of the twentieth century. Not only did this analysis bring to light the aforementioned inability on the part of classical physics to render a rational and coherent account of itself, but what is still more important, it led to the startling conclusion that no such account can be given at all.

---

7. Sir Arthur Eddington, *The Philosophy of Physical Science* (Ann Arbor, MI: University of Michigan Press, 1958), p32.

Now this impossibility derives from the existence of certain quantities within the classical scheme that turn out to be in principle unmeasurable, or better perhaps, to be measurable only with a limited degree of accuracy. As might be expected, the 'mesocosmic' range of physical reality within which classical physics had proved its worth coincides exactly with the range of physical magnitudes for which that 'limited degree of accuracy' is sufficient to preclude observable discrepancies. Outside the mesocosmic range, classical physics breaks down. To proceed beyond these limits, one requires theories in which at least one of the classical 'unobservables' has been successfully eliminated through the creation of a new mathematical formalism.

To indicate a little more concretely what we have been speaking of in very general terms, let us now consider the familiar concept of 'simultaneity'. We are normally quite certain that simultaneity is well defined on a global scale (as if the mere utterance of the magic word 'now' could suffice to determine a particular instant of time throughout the length and breadth of the universe!). Yet if we begin to ask ourselves by what kind of observations one could conceivably determine whether or not two widely separated events are 'simultaneous', we quickly discover that the matter is not quite so simple. Thus if a bolt of lightning should strike the front end of a moving train and another should strike the rear, it may turn out that these two events are simultaneous when observed relative to the train and yet not simultaneous when observed from the ground. What is more, the order of precedence (whether A precedes B or B precedes A) will also depend in general on our choice of reference frame. Of course, so long as we are dealing with a pair of events (A, B) which are not separated by vast astronomical distances and with two reference frames whose velocity with respect to one another is small compared to the speed of light, these discrepancies will not be observable. In other words, under ordinary conditions of measurement the notion of simultaneity retains an absolute significance. Outside of this restricted domain, on the other hand, the relativity of this concept—or equivalently, its 'unobservability' in absolute terms—comes into play. And when that happens classical physics breaks down.

Now as Einstein has shown through his 'special theory of relativity', the underlying difficulty can be resolved by fusing physical space and physical time together into a four-dimensional space-time, which in effect gets rid of the notion of absolute simultaneity. And as one knows, this theory has led to brilliant and most astonishing results (including the fateful formula $E = mc^2$). It has been accurately confirmed by innumerable measurements and observations, and has given rise to various remarkable technological developments. Moreover, it forms the starting point of a still more sophisticated theory (the gravitational and unified field theories), which might well be described as a fusion of space-time and matter into a so-called curved space-time, whereby matter—and even electromagnetic fields—are in effect reduced to 'geometric' properties of the underlying continuum.

It is to be noted that these relativistic theories reduce to classical physics in the mesocosmic range. Formally speaking, they reduce to the classical theory in the limit as the speed of light tends to infinity, which is just the limiting case in which distant simultaneity has a physical meaning. Relativity theory is therefore a refinement of classical physics, based upon the elimination of a particular 'unobservable'. Its range, moreover, extends far beyond the confines of the mesocosm into the world of astronomical dimensions: the physical macrocosm. On the other hand, the range of relativity theory, too, is not unlimited, due to the fact that it has taken over certain other classical unobservables, namely, quantities that become unobservable toward the opposite end of the scale: within the world of atoms and fundamental particles.

By way of illustration, let us consider the 'position-and-velocity' of a particle. Now according to the classical description, every particle or mass point has a well defined position and velocity at every instant of time. With respect to a local coordinate system, one thus obtains for every coordinate direction a pair (q, v) of position and velocity coordinates. As it turns out, (q, v) is a microcosmic unobservable. For indeed, according to the famous Heisenberg uncertainty principle, the greater the accuracy with which one is able to determine one of the two coordinates, the less one is able to know about the other. More precisely, if we replace the velocity v by the corresponding momentum coordinate $p = mv$ (where m denotes the mass), the principle asserts that the product of the respective uncertainties of q and p cannot be less than the so-called Planck's constant h. Needless to say, inasmuch as h is a very small quantity (approximately 6.626 x 10^{-27} erg sec), this unobservability of (q, p) does not manifest itself under the usual conditions within which measurements are carried out. When it comes to the observation of atoms and fundamental particles, on the other hand, it does show up, and in fact turns out to play a crucial role. And this is the reason why both Heisenberg and Schrödinger have devised a new mathematical formalism (the two were later found to be equivalent) that eliminates (q, p), along with a host of similar unobservables. The resultant theory, moreover, has at one stroke brought order into the chaos of earlier quantum mechanical theorizing, and it has been enormously successful in explaining a wide range of microphysical phenomena. At least within the first level (so to speak) of the microphysical domain, it could well be the 'right' theory. As might be expected, the new quantum mechanics reduces to the classical theory in the limit as h tends to zero (in the limiting case, that is, in which 'position-and-momentum' becomes observable).

IN A WAY, both relativity theory and quantum mechanics have 'desolidified' the physical universe. More precisely, they have demonstrated the insufficiency of those ordinary notions about 'matter' which derive partly from common sense, and partly from classical physics. While these conceptions may have a semblance of truth and a high degree of utility within the mesocosmic range, their validity is strictly limited to that domain. The mesocosm itself has thus been deprived of its seemingly absolute reality and been reduced to the status of a phenomenon. It has become an aspect of the physical universe in relation to man. Strictly speaking, we fall into illusion the moment we forget this relativity, and thus attribute to that 'cosmos' a kind of independent reality which it does not possess.

But what about the new physical theories: can these, perhaps, provide a more-than-phenomenal knowledge of the universe? In the case of relativity (which is actually a theory of invariants, that is to say, of quantities that are independent of particular observations), it is in some degree optional whether we wish to attribute a more-than-formal reality to the invariant structure, be it a curved space-time or something else. Expert opinion has been divided on this issue, and whereas Einstein himself inclined towards a realist interpretation of his theory, it seems that a majority of leading physicists do not share this view. To a large extent the answer will depend on how seriously one takes the new quantum mechanics. For indeed, that theory compels us to admit that our scientific knowledge is incurably phenomenal. It is a knowledge, in other words, not of things in themselves, but of things in relation to an observer. As Heisenberg has expressed it, 'if one can speak of a world-view (*Naturbild*) of the exact sciences in our time, this actually refers no longer to a view of Nature, but to a view of our relationships to Nature.'[8]

In the case of quantum mechanics, this subjectivity is reflected in its very formalism. In the Schrödinger formulation, the physical system is formally represented by a so-called wave-function, which however cannot be interpreted as a description of the physical system as such, but rather embodies our *knowledge* of that system. There has been a great deal of debate as to whether this knowledge is inherently statistical, so that the wave-function becomes a kind of 'catalogue of expectations', as Schrödinger has put it. In any case, it is in some way a 'catalogue of information' (to use Pauli's phrase), and this information can be extracted from it through the application of mathematical operators, which formally represent measurable quantities. There is an operator, for instance, representing the position coordinate q of a particle, and another operator representing the corresponding momentum coordinate p. But there is no operator representing the unobservable pair (q,p)! Moreover, an operator cannot in general extract a precise value from the 'catalogue', as this would obviously lead to an exact determination of unobservables (such as $[q,p]$, for instance). Furthermore, the mathematical formalism itself guarantees that the precision of the infor-

8. Werner Heisenberg, *Das Naturbild der heutigen Physik* (Hamburg: Rowohlt, 1955), p 21.

mation which any given wave-function contains with regard to the variable q, let us say, is inversely proportional to the precision of information relating to the so-called conjugate variable p. The fact is that the Heisenberg uncertainty principle can be derived from the formalism as a mathematical theorem.

The wave-function is thus a 'catalogue of information' which does not tell us all that we might wish to know about a given physical system. The point is, however, that it tells us all that we *can* know. This becomes somewhat plausible if one considers that every measurement involves a physical interaction between two systems: the system to be measured, and a second system, through which the measurement is to be accomplished (consisting of scientific instruments, plus light rays or other 'test particles'). Obviously, therefore, the measurement itself disturbs the first system to some degree. Now, quantum theory affirms that the transfer of energy between two systems is not an inherently continuous process, but involves discrete units or 'quanta', which have a small but fixed value. This implies, in particular, that the disturbance to the first system cannot be reduced beyond certain limits, if the measurement is to take place at all. The uncertainty principle may therefore be interpreted to mean that a measurement of q, let us say, disturbs the particle so as to affect a subsequent measurement of p. The sharper the measurement of q, the greater will be the disturbance to the observed particle and the resultant uncertainty of p.

On the other hand, this interpretation must not be pressed too far. It tacitly assumes that a particle in itself has a definite position and momentum, notwithstanding the fact that we may not be able to measure the one without perturbing the other in some uncontrollable way. But clearly, this assumption is unwarranted and unverifiable. It is the sort of thing that physics has been at great pains to discard for the past seventy or eighty years. One might almost say, it is the kind of hypothesis that carries a price on its head. To re-introduce it now, when it is not needed, and can tell us absolutely nothing, would be to have missed the point of what physics is all about.

But it happens that there is a still more compelling reason which forces us to abandon the concept in question: the presumed position and momentum do not exist because, strictly speaking, there is no such thing as a 'particle' in the first place. The point is that by virtue of the so-called wave-particle dualism, one can speak of particles only in relation to certain kinds of experiments, with the understanding that the same underlying physical reality will display itself as a continuously distributed wave in relation to other kinds of experiments. Inasmuch as the concepts of particle and wave exclude one another logically, one is forced to conclude that the physical reality *in itself* is neither particle nor wave. All that one can say is that in certain respects it acts as if it were a particle, while in other respects it acts as if it were a wave. Let us add that this wave-particle dualism applies to all forms of matter or energy, whether it be electromagnetic 'waves' (e.g., light) or fundamental 'particles' (e.g., electrons).

Obviously this remarkable fact places the uncertainty principle in a new perspective. Indeed, it demands such a principle. One might say that the uncertainty in question is just sufficient to prevent us from pinpointing the so-called particle to an extent that would rule out the wave characteristic. It thus provides just the necessary leeway that permits the wave-particle dualism to exist. What appears as a gap in our knowledge from a more or less classical point of view, turns out to be simply a measure of false expectations. Quantum theory gives nothing away. It does justice to both sides of the wave-particle duality without forfeiting one whit of possible knowledge. It is not quantum theory, therefore, which frustrates our classical demands, but Nature herself: the reality does not conform to the dream.

IT APPEARS THAT the classical description of the physical universe has eroded to the point where little, if anything, is left. For all its precision and quasi-absolute pretensions, that picture has proved to be 'human, all too human'.

On the other hand, one might reply that the Newtonian description, after all, was just a first shot at the target, an initial attempt to construct an adequate model of physical reality. With the progress of science one can naturally expect to arrive at better and better models, which in course of time will provide an ever more accurate picture of 'the way things are'.

But it turns out that this optimistic appraisal is fundamentally in error. Curiously enough, as the picture comes into sharper focus, so to speak, it shatters and eventually disappears. A point is inevitably reached where the picture itself dissolves, leaving only a set of field equations or an operational calculus as a kind of formal skeleton of what had once been a physical world-view. It seems that our quarry has mysteriously eluded the net just at the moment when it was about to be seized. For as Schrödinger has observed, it was precisely during the years or decades which let us succeed in tracing single, individual atoms or fundamental particles, that we have been compelled to dismiss the idea of such particles as 'fundamental entities'.[9] Thus, whatever it may be that has actually been seized (and speculations in that regard have covered a very broad spectrum), it is evidently not the original quarry: the Newtonian things-in-themselves, that were said to populate the physical universe.

But what, then, has become of the physical universe itself? What is the status of this idea in the light of contemporary knowledge? From a purely technical point of view, one sees readily enough that the concept plays no role whatsoever in the economy of exact scientific thought, nor has it ever done so in the past. And yet the idea remains with us as an implicitly assumed background, a mental presupposition that serves to shape and define the general scientific outlook. If it is to be admitted that the immediate object of scientific inquiry is what Heisenberg calls 'our relationships to Nature', then for all but the most astute, at any rate, that Nature is still what it has been ever since the days of Newton: in a word, it is the physical universe.

Now the fact that this notion—or equivalently, the bifurcation postulate—has proved to be devoid of scientific sanction does not in itself invalidate that concept, or that postulate: it only renders it optional, in a way, and curiously irrelevant to the business of science. Meanwhile the premise remains what it has always been: a

9. Erwin Schrödinger, *Science and Humanism* (Cambridge: Cambridge University Press, 1951), p17.

metaphysical assumption, which stands or falls on strictly philosophical grounds. It will be of interest, therefore, to go back once
more to the early beginnings, in order to recount the origin and
subsequent philosophical fortunes of this crucial idea.

2

THE CARTESIAN DILEMMA

As WE HAVE NOTED BEFORE, the idea of bifurcation began to take shape during the sixteenth and seventeenth centuries, and was associated from the outset with the formation of the new physics. Among the various factors which have influenced this development, the most important, it would seem, was the revival of Platonic scholarship, headed by such men as Marsilio Ficino (1433–99) and Pico della Mirandola (1463–94). Once again the ideas of number and harmony began to exert their perennial power to enthrall. We know that Nicolaus Copernicus (1473–1543) came under the direct influence of this school while he was a student at Bologna, and certainly the subsequent triumph of his astronomical theory could only serve, in turn, to strengthen an already growing enthusiasm for the mathematical sciences. With astonishing zeal, men began to look upon mathematics as the prototype and prerequisite of true knowledge, and quite possibly, as the only source of certitude. It appears that Kepler (1571–1630) was speaking for the entire age when he declared that 'just as the eye was made to see colors, and the ear to hear sounds, so the human mind was made to understand, not whatever you please, but quantity.'[1]

Unmistakably, a transition from the medieval to the modern cast of mind is in full progress. It is evident that the stage is being set for the Newtonian discoveries; and yet Kepler himself is still visibly imbued with the transcendental leanings of Platonism, and it is hardly an accident that his scientific interest remains fixed upon the sun and its planets. One feels that the true object of his quest was

1. *Joannis Kepleri Astronomi Opera Omnia* (Frankfurt and Erlangen, 1858), I, 31; cited in Burtt (n2 above), p57.

not a matter of empirical laws and correlations, but the discovery of eternal harmonies.

With Galileo (1564–1642) the direction and focus of the scientific gaze begins to shift noticeably: from heaven towards the earth, as one might say. The Tuscan scientist continues to extol the pre-eminent virtues of mathematics, and even inveighs occasionally against the fickle and illusory nature of sense knowledge. But while he takes over these Platonic themes, he begins to bend his energies to the accomplishment of a very un-Platonic task: the mathematical explication of such mundane things as falling stones. At the same time, he is coming visibly under the influence of another idea, which was somehow gaining a hold upon the European mind: the idea of mechanism. As historians of science have pointed out, this conception was already beginning to express itself during the fourteenth century in the form of a remarkable craze for the construction of gigantic astronomical clocks. 'No European community felt able to hold up its head unless in its midst the planets wheeled in cycles and epicycles, while angels trumpeted, cocks crew, and apostles, kings and prophets marched and counter-marched at the booming of the hours.'[2] It may well be that these prodigies of mechanical art suggested the idea that celestial motions and other natural phenomena could be somehow accounted for in mechanical terms. In any case, by the seventeenth century the concept of a 'clockwork universe' had become very much a part of the European intellectual scene, and was exercising a considerable scientific influence. One might add in passing that the intimacy of this connection between 'mechanism', in the sense of physics, and 'clockwork', is further illustrated by the fact that the mechanical discoveries of Galileo were almost immediately incorporated into the construction of a pendulum clock, invented by Huygens in 1656. But whatever may have been the source of the idea, it is evident that the concept of mechanism went hand in glove with the prevailing penchant for mathematics, and contributed one of the essential ingredients of the new Weltanschauung. One thing more was needed, and that is the bifurcation

2. Lynn White, *Medieval Technology and Social Change* (Oxford: Oxford University Press, 1962), p124.

postulate. Unobtrusively, and presumably without realizing the enormity of this step, Galileo came forth to supply the missing element by enunciating a subjective interpretation of the so-called secondary qualities.

It was René Descartes (1596–1650), however, who gave the new vision its fully articulated form. The French mathematician, physicist, and philosopher, fired by the same influences and dreams as his Italian compeer, brings to the scene a powerful metaphysical bent of mind. He too perceives mathematics as the essential instrument of scientific knowledge, and is passionately devoted to the cause of universal mechanics. He strives to lay the theoretical foundations for a rigorous mechanical science, based upon mathematical principles which would be able to explain the workings of Nature, from the movement of planets to the fine motions associated with animal bodies. But he also understands well enough that only a mechanical universe can be comprehended in mechanical terms. The point is made forcefully in the following passage:

> We can easily conceive how the motion of one body can be caused by that of another, and diversified by the size, figure and situation of its parts, but we are wholly unable to conceive how these same things (size, figure and motion), can produce something else of a nature entirely different from themselves, as, for example, those substantial forms and real qualities which many philosophers suppose to be in bodies. . . .'[3]

Thus, with remarkable acumen, he observes that 'those substantial forms and real qualities which many philosophers suppose to be in bodies' could not be accounted for in mechanical terms. It is clear to him, in other words, that the possibility of universal mechanics hinges upon bifurcation. Somehow the secondary qualities (such as color and sound) need to be eliminated from the objective world, and Descartes presumes to accomplish this through what is nowadays referred to as the Cartesian mind-body dualism.

3. *Principia philosophiae*, in *Oeuvres* (Paris, 1824), iv, 198; cited in Burtt (n2 above), p112.

We need not attempt to follow Descartes in his solitary meditations, wherein he sought to touch the bedrock of human knowledge. Suffice it to say that he emerged from his garden retreat fully convinced that the universe is precisely what it must in fact be, if it is to submit to mechanical description. In short, it is a mechanical world, made up entirely of *res extensa* (the later Newtonian 'matter'), moving in space according to mechanical laws. All the rest is to be relegated to *res cogitans* or thinking substance, which exists in its own right as a kind of spiritual entity. It is noteworthy that Descartes came to this *res cogitans* at the outset of his meditations through the famous *cogito ergo sum*. It appeared to him as the one and only immediate certainty, whereas the existence of a mechanical universe, external to the *res cogitans*, was to be arrived at later through a logical argument, in which the idea of God and His veracity plays the leading role. It is indeed a remarkable irony that the basic premise of modern materialism should initially have been founded upon theology!

By and large, Descartes was sensitive to the enormous philosophical difficulties raised by the dichotomy of *res extensa* and *res cogitans*. In the first place, if the *res cogitans* has no extension, how can the *res extensa* act upon it, as it is presumed to do in sense perception? And how can presumed motions within a human brain give rise to unextended conceptions of an extended universe? Or conversely, how can *res cogitans* influence the motion of *res extensa* in the case of volitive action? If we are 'wholly unable to conceive' how mechanical causes could produce 'those substantial forms and real qualities which many philosophers suppose to be in bodies,' how then are we to conceive the interplay of *res cogitans* and *res extensa*? Like Galileo and others of his time, Descartes is sometimes willing to resolve philosophical difficulties by recourse to Deity. By means of pseudo-theological arguments (singularly unconvincing to materialist and believer alike), he seeks to extricate himself from a philosophical impasse, brought on by his own postulates. At other times, however, he seems to forget the impasse, and speaks as if there were really no problem at all. The following passage, for instance, reveals Descartes in one of these confident moods:

But since we know, from the nature of our soul, that the diverse motions of body are sufficient to produce in it all the sensations which it has, and since we learn from experience that several of its sensations are in reality caused by such motions, while we do not discover that anything besides these motions ever passes from the organs of the external senses to the brain, we have reason to conclude that we in no way likewise apprehend that in external objects which we call light, color, smell, taste, sound, heat or cold, and the other tactile qualities, or that which we call their substantial forms, unless as the various dispositions of those objects which have the power of moving our nerves in various ways. . . .[4]

But how is it ever possible that mechanical causes ('the various dispositions' of the perceived objects) could give rise to sensations such as redness? Is this not to assert, once again, that such things as size, figure, and motion 'can produce something else of a nature entirely different from themselves'?

For better or for worse, this is the philosophical legacy which Descartes passed on to Newton, who in his turn transmitted it to the scientific world at large. It is to be observed that ere long, scientists accepted the *res extensa* as a gospel truth, while totally rejecting the arguments by which Descartes had sought to bolster that notion. At the hands of the British school, moreover, the *res cogitans* (which had originally been conceived as an unextended substance) was at first imprisoned within a ventricle of the brain (the so-called Newtonian sensorium), and later eliminated *in toto*. By a curious reversal of Cartesian logic, the *res extensa* gained precedence over the *res cogitans*, or as one might almost say: the conjecture swallowed up the dream.

BY THE TIME the seventeenth century drew to its close, the idea of a mechanical universe was rapidly gaining ground as the official

---

4. *Principia*, IV, 199; cited in Burtt (n2 above), p112.

doctrine of science. It seems that for all but a discriminating few, consisting mainly of philosophers, every new triumph of physics was counted as yet another incontrovertible vindication of the Newtonian world-view. For their part, the men of science—far more interested in extending the boundaries of victorious analysis than in scrutinizing its foundations—were in no mood to question this argument. Generally speaking, it was an age of incredible optimism.

But there were also some rugged intellectual individualists who refused to conform to the trend. In the year 1710, for instance, one encounters George Berkeley, a spirited and eloquent Irishman, propounding arguments of great force against the Cartesian concept of an unperceived and unperceivable universe:

> The table I write on I say exists; that is, I see and feel it: and if I were out of my study I should say it existed; meaning thereby that if I was in my study I might perceive it, or that some other spirit actually does perceive. . . . For as to what is said of the absolute existence of unthinking things, without any relation to their being perceived, that is to me perfectly unintelligible. Their *esse* is *percipi*; nor is it possible they should have any existence out of the minds or thinking things which perceive them.[5]

Such arguments are not easily countered. Clearly, they attack Cartesianism in its most vulnerable spot, and with its own weapon, one might add. For 'some truths there are so near and obvious to the mind,' writes Berkeley in unmistakably Cartesian strains, 'that a man need only open his eyes to see them.' Yet, surprisingly, what the two men see is altogether different! In place of a mechanistic universe, existing by itself in a perpetual isolation which no eye has ever pierced, the Irish bishop beholds a world of color, sound, and fragrance, whose essence it is to be perceived. He too has meditated upon the foundations of human knowledge, only to be convinced 'that all the choir of heaven and furniture of the earth, in a word, all these bodies which compose the mighty frame of the world, have

---

5. *Principles of Human Knowledge*, I, 3.

not any subsistence without a mind'; and finally, that 'there is not any other substance than *Spirit*, or that which perceives.'[6]

Seventy-one years after the first publication of Berkeley's *Principles* the center of debate shifted abruptly to the small German town of Königsberg, where a sedate and meticulous professor astonished the world with a ponderous dissertation: the *Kritik der reinen Vernunft.* Like Descartes, Kant also was concerned to place the science of mechanics upon a firm theoretical foundation. He had listened carefully to the ongoing philosophical controversy, and understood quite clearly that the crux of the difficulty resided in an impassable gulf which separated the scientist from his objects. Kant's solution to the problem, basically, was to bring the objects over to the hither-side of the gulf. With remarkable acumen, the Prussian philosopher sets out to establish his position. As his opening gambit, he observes that 'by means of outer sense, a property of our mind, we represent to ourselves objects as outside of us, and all without exception in space.'[7] With relentless logic, he proceeds to unfold the contents of his premise:

> Space is not an empirical concept which has been derived from outer experiences. For in order that certain sensations be referred to something outside me (that is, to something in another region of space from that in which I find myself), and similarly in order that I may be able to represent them as outside and alongside one another, and accordingly as not only different but as in different places, the representation of space must be presupposed. The representation of space cannot, therefore, be empirically obtained from the relations of outer appearance. On the contrary, this outer experience is itself possible at all only through that representation.[8]

Undeterred by the astonishing nature of his claims, Kant drives on to the inevitable conclusion of his argument: 'Space does not

---

6. Ibid., 1, 6, 7.
7. *Critique of Pure Reason* (New York: Random House, 1958), p 43.
8. Ibid., p 43.

represent any property of things in themselves, nor does it represent them in their relation to one another.'[9] It belongs to the world of appearances as a form superimposed, so to speak, by our intuition (*Anschauung*).

This inquiry is immediately followed by another: the analysis of time. If space has been found to be the 'pure form of outer intuition', time turns out to be the 'pure form of inner intuition', and indeed, 'the formal *a priori* condition for all appearances whatsoever.' The position is summarized in the following terms:

> What we have meant to say is that all our intuition is nothing but the representation of appearance; that the things which we intuit are not in themselves as they appear to us, and that if the subject, or even only the subjective constitution of the senses in general, be removed, the whole constitution and all the relations of objects in space and time, nay space and time themselves, would vanish. As appearances, they cannot exist in themselves, but only in us. What objects may be in themselves, and apart from all this receptivity of our sensibility, remains completely unknown to us. We know nothing but our mode of perceiving them. . . .[10]

We need not pursue the Kantian argument beyond this point. Bold and provocative, it puts everything in a brand new perspective. In its own way, it resolves the bifurcation impasse, and provides a conceivable basis for a rigorous justification of scientific knowledge. Moreover, one can hardly fail to observe that the Kantian perspective bears special relevance to the situation of physics today: to the idea, that is, of a science whose true object is 'our relationships to Nature'. Although there is little reason to suppose that the scientific world at large has ever paid the slightest attention to the Königsberg philosopher, it is certain that twentieth-century physics owes much to his penetrating criticism of the Newtonian foundations. The fact is that European philosophy was never again the same. If Hume had

9. Ibid., p 46
10. Ibid., p 54.

awakened Kant from his 'dogmatic slumbers' (as we know by his own admission), Kant himself has had a similar effect upon succeeding generations of thinkers.

EVEN SO, the spell of Cartesianism was not yet broken. In retrospect, it appears that until the beginning of the twentieth century, the major schools of Western philosophy continued to labor under the burden of a certain Cartesian prejudice. Despite their vigorous criticism of the French savant, not only Berkeley and Kant, but others as well, had unwittingly taken over his central premise. For well over two centuries, this veritable *idée fixe* retained a kind of strangle-hold upon European philosophy, which few thinkers, if any, were able to break.

Basically, this Cartesian premise reduces to the belief that the true object of sense perception must somehow reside within the confines of the human mind. More precisely, it affirms that perception does not in fact transcend what is immediately given in the form of sense data or mentally constructed images derived therefrom. Now this is the assumption that renders the so-called external world unperceived and unperceivable. Thus, if such a universe exists at all, it is in any case conjectural. In other words, it becomes a thing-in-itself, whose existence can be doubted *à la* Descartes, or denied *à la* Berkeley. At the same time the familiar world, as given in the ordinary human experience, is rendered subjective and in a way unreal: it becomes essentially a private phantasm, the kind of thing whose *esse* is *percipi*, as Berkeley had well observed. One cannot but agree with him that it would be 'plainly repugnant' to suppose that such entities could exist 'outside the minds or thinking things which perceive them.'

Once the Cartesian premise has been assumed (and not before!), bifurcation becomes a conceptual possibility. One is then free to conceive of an external universe bereft of all but mechanical properties: the common-sense objection that the world quite obviously does not conform to that description has now been deprived of its force. At one stroke, the objective world has become an unknown entity, to be somehow unveiled through the cogitations of the philosopher, or through the scientific inquiries of the physicist. But this possibility itself becomes dubious. The more carefully one investigates, the more it appears that the gulf between the external realm and its subjective representations is in fact impassable. It was not without reason, therefore, that Bishop Berkeley denied the existence of the external world. It is to be noted, however, that his entire argument rests squarely upon the Cartesian premise. In effect, the Irish philosopher has demonstrated, quite conclusively, that if perception terminates in a mental image, then it follows that the notion of an external universe is inherently self-contradictory. One should add that even the philosophical revolution inaugurated by Kant has failed to resolve the fundamental difficulty. The gulf remains, and is all the more sharply delineated through the precisions of the Kantian analysis. It is no longer enough to say that the external universe is unperceived and unperceivable; for in the form of the Kantian *Ding an sich*, it has lost not only its 'secondary' but also its 'primary' attributes. Ghost-like, it remains as the ultimate unknowable X around which the human mind has fabricated the known and knowable world.

It was not until the beginning of the twentieth century that the Cartesian premise became the target of serious philosophic criticism. Since then, however, the recognition has become widespread among philosophers that the true object of perception does not in fact reduce to a mental image. There is that which is passively received (the datum), and there is that which is seized through an act of intelligence: call it the object of intentionality. It is no doubt true that the intentional act entails a complex process, involving intermediate mental representations. Yet what is actually perceived is not the sense datum, nor any subjective representation or image, but quite simply the intentional object. Now to say that this object, as the terminus of an intentional act, must again be an appearance, or a subjective representation—that is surely an assumption. In fact, it is precisely the Cartesian premise! Admittedly, this premise seems quite plausible so long as we implicitly submit to its claim. Yet, viewed from neutral ground, it immediately becomes suspect. Thus, in the case of visual perception, for example, the intentional object is evidently three-dimensional, a circumstance which in itself suggests rather strongly that the object in question is not just a visual image. To maintain, moreover, that it must nonetheless be a subjective representation (a thing whose *esse* is *percipi*) is to take a totally unwarranted step. It is to assume from the outset that 'the soul has no windows', and perhaps, in the final analysis, to postulate the impossibility of objective knowledge as such.

THESE OBSERVATIONS relating to the nature of perception, and of intentionality in general, are intended, not to close an argument, but rather to introduce the basic issue. The problem is admittedly difficult, and far more weighty than might appear at first sight. It calls for the most careful consideration. Moreover, it has been the subject of painstaking investigation on the part of leading philosophers, beginning with Edmund Husserl (whose studies were initiated around the beginning of the twentieth century).

A mathematician by training, Husserl began his philosophical inquiries with a keen analysis of purely logical conceptions relating to the foundations of mathematics. With remarkable cogency he defended the objectivity of these logical objects against subjectivist doubts, and it would appear that in the process he succeeded in establishing the transcendence of certain intentional acts. Later Husserl extended the scope of these investigations to other modes of intentionality, while forging a general philosophic method for the accomplishment of such analyses. So far as the bifurcation tenet is concerned, he claims to have established the objective character of numerous types of intentional objects, including the familiar entities of sense perception. Yet be that as it may, it is at least certain that the powerful lenses of Husserl's 'phenomenological analysis' have brought to light the insufficiency of the Cartesian conceptions.

Before long another outstanding figure had joined the fray against residual Cartesianism: it was another mathematician turned philosopher, namely Alfred North Whitehead. He too had occupied himself early in his career with foundational questions relating to mathematics, to the extent of becoming one of the founders of mathematical logic. Endowed with a broad scientific background and a deep grasp of the new physics, he later turned his attention to the foundations of physical science. With exceptional clarity he came to perceive that these foundations had fallen into an advanced state of philosophic disarray. Here is a typical passage from one of his many lectures, wherein he summarizes the contemporary situation:

> The state of modern thought is that every single item in this general doctrine [Whitehead is referring to the Newtonian scheme] is denied, but that the general conclusions from the doctrine as a whole are tenaciously retained. The result is a complete muddle in scientific thought, in philosophic cosmology, and in epistemology. But any doctrine which does not implicitly presuppose this point of view is assailed as unintelligible.[11]

At the same time, Whitehead understands full well the causes that have led to this impasse. As the following quotation will show, he perceives the merits of the Newtonian scheme, and the obstacles which stand in the way of its replacement. Yet he is no less cognizant of its irremediable limitations:

> In the first place, we must note its astounding efficiency as a system of concepts for the organization of scientific research. In this respect, it is fully worthy of the genius of the century which produced it. It has held its own as the guiding principle of scientific studies ever since. It is still reigning. Every university in the world organizes itself in accordance with it. No alternative system of organizing the pursuit of scientific truth has been suggested. It is not only reigning, but it is without a rival. And yet—it is quite unbelievable. This conception of the universe is surely framed in terms of high abstractions, and the paradox only arises because we have mistaken our abstraction for concrete realities.[12]

Here we have come upon one of Whitehead's key points: the so-called fallacy of misplaced concreteness. Repeatedly he brings home the idea that physical science is habitually confounding its 'high

---

11. *Nature and Life* (New York: Greenwood, 1968), p 6.
12. *Science and the Modern World* (New York: Macmillan, 1953), pp 54–55.

abstractions' with the primary reality. Thus one begins by abstracting from concrete existence, and ends by attributing concreteness to the abstraction. Or equivalently, one first cuts asunder what in truth is one, and then attributes an independent reality to one of the resultant fragments. But of course the error does not affect the reality: it only creates blindness. And so the scientific world-view entails a certain incomprehension, to which we have become habituated through an extensive process of indoctrination:

> Science can find no individual enjoyment in Nature; science can find no aim in Nature; science can find no creativity in Nature; it finds mere rules of succession. These negations are true of natural science. They are inherent in its methodology. The reason for this blindness of physical science lies in the fact that such science only deals with half the evidence provided by human experience. . . . The disastrous separation of body and mind which has been foisted on European thought by Descartes is responsible for this blindness.[13]

Summing up, one can say that Husserl and Whitehead have been the leading figures in the contemporary philosophical refutation of the Cartesian premise. At long last it appears that a definitive verdict in the philosophic trial of 'scientific' cosmology has been cast: its basic assumption has proved to be untenable.

In a way, this marks a return to the natural and unperverted conceptions of mankind. In spite of the learned debates, the Cartesian premise had always remained *de facto* unbelievable, and perhaps it is precisely the opposing tenet—the naive and commonsense view that we do look out upon the real objective world—that constitutes a truth 'so near and obvious to the mind. . . .' But be that as it may, it must not be supposed that the epistemological enigma has now been resolved, or that the problem perhaps had never really existed. For indeed, to recognize (as Husserl and others have done) that perception transcends the subjective domain is hardly to explain how this prodigy can be accomplished. The fact is that the basic riddle

13. *Nature and Life*, p30.

remains—if only we have an eye to see it!—and one should add in fairness to Descartes and his successors, that when they erred, it was not over trifles.

ON THE WHOLE, the later philosophical development (which we have sketched) has had little direct impact upon the scientific mentality of our time. Despite the breakdown of classical physics, the Newtonian metaphysics is still in force, and so is the Newtonian positivism. Now as before, it can be said that men of science have learned their philosophy from the *Principia*. They have learned it so well, in fact, that these modes of thought have become ingrained to the point of conferring upon the Newtonian philosophical premises a status of self-evidence (which is no doubt the reason why 'any doctrine which does not implicitly presuppose this point of view is assailed as unintelligible'). There have been some notable exceptions, of course, as we have had occasion to point out. Yet, by and large, the scientific mentality has remained impervious to post-Newtonian philosophic influence. It is plainly evident that neither Kant nor Whitehead has yet succeeded in rousing the scientific community at large from its 'dogmatic slumbers'. As Whitehead himself has pointed out, the general conclusions of the Newtonian doctrine have been 'tenaciously retained', and so far as the resultant 'muddle' is concerned, it would appear that few scientists have been unduly disturbed.

On the other hand, we have all become vastly more sophisticated, and in certain respects our basic conceptions about science have changed. For example, we are beginning to sense that the scientist is more than a mere spectator. We tend to see him as contributing an element of creativity to the scientific process, and lending shape to our scientific knowledge. To some degree, we have come to perceive physics as the interplay of external Nature with the devices and strategies of the physicist. In line with this trend, the idea of 'models' has been steadily gaining ground. The recognition has begun to take hold that science deals, not with the physical world *per se*, but with various theories, each of which embodies certain aspects of truth. Looking back upon Newtonian physics, we now perceive that for all its brilliant success, it too is just a particular theory, and not a quasi-absolute, as one had previously supposed. It is one model among many others, each of which has its use and its inherent limitations. Hardly anything pertaining to physical theory is still regarded as sacrosanct. The scientific mentality has become at least partially pragmatic, and somewhat less prone than before to idolize its own creations. The very concept of 'model' implies a certain awareness of limitations, a falling short of absolute or complete knowledge, if not an element of relativity and the likelihood of being superseded.

Even so, this newly acquired sophistication does not in itself provide enlightenment on fundamental issues, nor does it remove the 'muddle in scientific thought, in philosophic cosmology, and in epistemology' to which we have previously alluded. In a way it serves to promote a climate of superficiality—a facile pluralism—which obviates and conceals the basic problem, instead of resolving it. 'Of course,' says Whitehead, 'it is always possible to work oneself into a state of complete contentment with an ultimate irrationality.'[14] Such an attitude goes hand-in-hand with pragmatism, or with what Whitehead refers to as 'the popular positivistic philosophy,' for in effect this perspective replaces truth by the notion of utility (generally conceived in narrow and rather primitive terms). One might even surmise that this outlook aims precisely at such a 'state of complete contentment,' be it with ultimate irrationality, or with anything else.

But clearly, these are matters which pertain more to the psychology of science than to its logical content, which is our primary concern. The fact remains that science does promulgate a doctrine. It makes claims about the nature of the physical universe which have a fundamental bearing upon other spheres of thought. Whether directly or indirectly, it inculcates certain metaphysical beliefs, and predisposes against others. Moreover, it addresses itself not only to scientific specialists, but to mankind at large. It has something very general to say about the world and our place in the world. In short, it has a truth to proclaim, a truth which according to official belief is

14. Ibid., p23.

founded upon hard and incontrovertible discoveries. This was the case during the classical or Newtonian era, and it is still the case today. Our contemporary sophistication and pragmatic propensities do not alter this fact: they only obscure it in some degree. Vast claims have been made, which need to be carefully analyzed, and ultimately judged.

Basically, our scientific world-view remains what it has been from the start; it is not the foundations, but only the superstructure that has changed. To be sure, physics has undergone a stupendous development, beginning from the rudimentary content of Newton's mechanics. It has become enriched, step by step, through the addition of new disciplines (such as the magnificent theory of electromagnetic fields), and after passing through various dramatic upheavals, has penetrated, on the one hand, into the mysterious world of fundamental particles, and on the other, into the far reaches of the galactic universe. In recent decades, moreover, it has even brought to birth a new scientific cosmology that claims all of space, time, and matter as its proper domain. And yet, as a world-view, this vast body of physical theory still rests upon the old Newtonian foundations. In point of its most essential content—which perhaps also constitutes the most gigantic of all its claims—it reduces, now as before, to the venerable Cartesian doctrine. Thus, in spite of all that has transpired over the past three-and-a-half centuries, that much-disputed hypothesis still constitutes the metaphysical foundation of modern science, implied (as we have seen) in the very concept of the physical universe.

On the other hand, it might also be argued that this concept proves, after all, to be irrelevant to physics on a technical plane, and so amounts to little more than a private fancy, bereft of scientific sanction. In this perspective, physics has no metaphysical foundation at all, nor does it require any such premises. For when it comes to the actual *modus operandi* of physics, we are dealing, not with the physical universe, but with things defined in terms of concrete procedures that have nothing whatsoever to do with metaphysical speculations. That is what positivism or operationism has always maintained—we encounter this position already in the *Principia*—and in a sense, it is perfectly true. Only we must also realize that this way of looking at the matter is not, strictly speaking, a world-view at all. It is rather a program of action, or if you will, it is the Weltanschauung of a computer. Nor does it seem likely that anyone could ever become so sophisticated—or so dehumanized—as to maintain a strictly positivistic perspective, without admixture of any metaphysically based notions. Yet be that as it may, when it comes to that broad and widely disseminated syndrome of beliefs which we have all along referred to collectively as the scientific Weltanschauung, it is quite evident that positivistic ideas can represent no more than one particular strand or level of thought. This is already clear from the fact that authentically operational definitions are accessible to none but the scientific expert, which implies that if our scientific outlook were formulated simply in such terms, it could never be popularized, or disseminated within wider groups. But most important of all, it would not be what it obviously purports to be: namely, a view of the real world, or more precisely, a doctrine concerning the nature of the physical universe. And so, with all due regard for the just claims of positivism, one must concede that when it comes to the scientific world-view, the concept of the physical universe has not by any means been displaced through operational notions.

Admittedly, this residual Cartesianism, which even today proves to be the fundamental ingredient of our scientific Weltanschauung, has been severely criticized since the twentieth century, and roundly condemned by some of the foremost thinkers. What is more, the attempt has been made, quite often, to construct a new theoretical foundation that might replace the classical scheme. Whitehead, for one, has brought forth a metaphysical doctrine that not only claims to resolve the Cartesian impasse, but purports to provide a new basis in terms of which the positive findings of science may be integrated into a coherent world-view. But whether or not Whitehead, or anyone else, has yet succeeded in this enterprise, the fact remains that these theories are understood and appreciated only within highly restricted circles. Invariably such speculations turn out to be quite technical, and far too difficult to commend themselves to a broader public. We should not forget, moreover, that the scientific community at large has so far shown little indication of being in any way dissatisfied with the metaphysical status quo, and for the most part has failed to grasp that there is really a problem in the first place. As we have noted before, a variety of factors have conspired to promote a type of mentality which almost instinctively shies away from the deeper questions. In such an intellectual climate the Cartesian confusion is bound to survive, undetected and unmolested by any rigorous inquiry.

This brings us at last to the seemingly paradoxical conclusion that the world-view associated with the most exact science is shot through with fundamental misconceptions. What we have collectively failed to grasp is that this purportedly scientific Weltanschauung is based, not upon the legitimate findings of science, but upon hidden philosophical or *a priori* assumptions which turn out in the last analysis to be self-contradictory. In the name of physics civilization has succumbed to a fantasy.

3

LOST HORIZONS

HAVING CONSIDERED THE QUANDARIES into which Western thought
has fallen under the influence of Cartesian philosophy, it behooves
us to re-examine the medieval position in its cosmological implications. What are the fundamental ideas, we must ask ourselves,
which distinguished the Christian world-view from the Cartesian
and post-Cartesian?

To begin with, it is to be noted that the modern conception of a
self-contained and self-sufficient universe is certainly incompatible
with the metaphysical teachings of Christianity. It is not enough to
say that the cosmos has been created by God and to maintain that
thenceforth it exists by itself and moves by its own energies and in
accordance with its own laws: surely the relationship between God
and world is far more subtle than that! One may put it this way:
God is not only transcendent, but He is immanent as well. Thus, on
the one hand, God transcends the cosmos: He resides beyond the
confines of space, 'in unapproachable light' as St Paul declares; and
yet, at the same time, He abides in all places and penetrates into the
innermost recesses of every existent thing. If God did not abide in
the cosmos, moreover, the cosmos would forthwith cease to exist.
As St Thomas Aquinas observes, 'since God is the universal cause of
all being, in whatever region being can be found, there must be the
divine presence.'[1]

Now this is just what the founders of modern science had failed
to grasp. It is not that they were atheists: did not Descartes, for one,
go so far as to found his belief in the existence of an external world

---

1. *Summa Contra Gentiles*, III, 68.

upon the presumed veracity of the Creator? And did not Newton devote the later years of his life to theological speculation? But despite their belief in the existence of God these men were estranged from the idea of divine immanence: theirs was a purely transcendent God, a mere Creator who had no further role to play and was not needed any more.

Perhaps, due to the mounting rationalism of the age, these thinkers found it difficult to resolve the seeming antinomy between the concepts of transcendence and immanence. In traditional parlance, they were caught between the horns of a theological dilemma: if one accepts immanence but rejects transcendence, one falls into the heresy of pantheism; and if, on the other hand, one accepts transcendence but rejects immanence, one falls victim to deism. Now the way of Christian orthodoxy lies neither to the right nor to the left but straight down the middle, 'between the horns'. In other words, it lies in the recognition that the antinomy is only apparent: a mere reflection of human incapacity, one could say. And indeed, all the basic truths of theology do in fact assume an antinomial appearance when formulated in dogmatic terms, beginning with the Trinitarian doctrine. If even an electron can manage to be both particle and wave, why should God Himself be constrained by what appears to us as an irreconcilable opposition? So, too, when it comes to the concepts of transcendence and immanence, Christianity counsels us to stand firm. It insists that each of these affirmations has something essential to say regarding the nature or action of God, and that both are indispensable for a correct understanding of the integral truth.

With the first signs of the Renaissance, however, this truth began to fade from the horizon of Western thought. As we have noted earlier, the age was visibly falling under the spell of the machine metaphor, a concept which evidently excludes the idea of divine immanence. In compliance with this metaphor, moreover, the founders of the new science were prone to assume that the world had been created somewhat in the manner of a clock, which after it has once been made and set in motion runs on by itself, and has no further need of its maker. To be sure, Newton himself did have some scruples on that score and thought it necessary that the Maker of the cosmic clock should intervene in the mechanism from time to time to set things right—an admittedly incongruous notion, for which also the great scientist was severely ridiculed by Huygens. But by the time Laplace had succeeded in demonstrating the dynamic stability of the solar system, if not before, it seems that these lingering doubts had been resolved to everyone's satisfaction.

By now men had become thoroughly accustomed to the idea that the universe consists of nothing but minute particles, and that every action—from the movement of stars and planets to the fine processes of life—is rigidly determined by mechanical laws. Under such auspices, moreover, the very concept of divinity cannot but appear strange and suspect, not to say useless; and it is no wonder, therefore, that before long the idea of a purely transcendent God—the God of Descartes and Newton—was given up too, at least as a subject for serious thought. As Laplace said to Napoleon when asked whether he believed in God: 'Sir, I have no need for that hypothesis.'

One might add that basically this position is nothing new. Since ancient times there have been materialists of that general description, even as there have been philosophers wise enough to understand that their position is flawed. In the words of Plotinus, 'those to whom existence comes about by chance and automatic action and is held together by material forces [could one imagine a more succinct description of the Newtonian scheme?] have drifted far from God and the concept of unity.'[2] Now what is this 'concept of unity' from which the materialists have drifted far? As we shall have occasion to see more clearly in what follows, it is tantamount to the immanence of God, the metaphysical fact that 'in whatever region being may be found, there must be the divine presence.'

NOT ONLY DOES GOD ABIDE in all places, but He reveals Himself through all beings. '*The heavens declare the glory of God and the firmament sheweth His handywork.*' Now this is not just poetry, in the contemporary sense. There was a time when men actually thought that the creation bears the imprint of its Maker, and that the cosmos

2. *The Enneads*, S. MacKenna, trans. (London: Faber & Faber 1930), VI, 9, 5.

mysteriously reflects the Face of God. Strange as it might seem to us, they believed that not only the stars and planets, but all the natural things of earth as well speak somehow of God as of a mystery, a secret to be intimated or half revealed. In a word, they surmised that the cosmos is a theophany, a manifestation of God.

While this conception can no doubt be found in the major metaphysical traditions of antiquity, it is especially germane to Christianity: we need but to recall the fact that Christian teaching is based upon the doctrine of the Logos, the Word of God, a term which in itself clearly suggests the idea of theophany. Moreover, what is implicit in the famous Prologue of St John is openly affirmed by St Paul when he declares that '*the invisible things of Him from the creation of the world have been clearly seen, being understood by the things that are made, even His eternal power and Godhead*' (Rom. 1:20). Could the notion of theophany—the idea that the creation manifests God—have been expressed with any greater clarity? And does not the Apostle allude elsewhere to the same vision when he tells us that '*now you see in a glass, darkly*' (1 Cor. 13:12)? As St Bonaventure explains clearly, 'the whole world is but a glass, full of lights manifesting the divine wisdom.'[3] The indisputable fact is that on its deepest level Christianity perceives the cosmos as a self-revelation of God.

Now it is true that in the development of theology the cosmological implications of the Christian legacy have been largely relegated to the background due to an overwhelming preoccupation with the soteriological content of the teaching. So far as the salvation of man is concerned, the decisive fact—from a Christian point of vantage—is not the cosmic theophany as such, but the self-revelation of God that began in Old Testament times and was consummated when *the Word became flesh and dwelt among us.* Yet, despite an understandable emphasis upon all that pertains most directly to the religious interests of man, the cosmological implications of the Christian revelation have not gone unnoticed. Like other aspects of the integral teaching, this too has found its place in the unfolding of theological thought.

3. *Collationes in Hexaemeron*, 11, 27.

But that is not all. So far from being simply a matter of speculative concern to theologians, the notion of cosmic theophany was implicit in a common Christian Weltanschauung in which all men, from learned doctors to the simplest peasant, could participate in some degree. In a very real sense the idea was part and parcel of our living cultural heritage right up to the beginnings of the modern age. As Sherwood Taylor maintains, 'before the separation of science and the acceptance of it as the sole valid way of apprehending Nature, the vision of God in Nature seems to have been the normal way of viewing the world, nor could it have been marked as an exceptional experience.'[4] Be that as it may, with the decline of the Middle Ages this 'vision of God in Nature' did become more and more exceptional, almost to the point of disappearing entirely from Western society. The world was fast becoming opaque, so to speak, and desacralized. A profound transformation of the collective consciousness was in evidence on all sides and in every sphere of culture. And whatever might be said for or against this metamorphosis, the fact remains that it represents a wholesale apostasy from the Christian world-view.

Already in the fourteenth century the spiritual culture of Europe was beginning to experience a decline. In the schools of theology, for instance, one encounters a marked formalist trend, a tendency to replace intellectual vision or spiritual contemplation by the workings of a formal methodological apparatus. The Baconian conception of a scientific method—a 'machine for the mind'—was perhaps already beginning to rear its head. In any case, the fruitful balance between vision and abstract thought—spirit and letter— which had led Latin Christianity into the golden age of Scholasticism has turned out to be precarious and short-lived. No sooner had the great masters passed from the scene than the schools began to manifest anti-metaphysical tendencies as well as certain signs of decadence. It appears that Europe was starting to lose its spiritual sight. And as the metaphysical vision waned, the modern scientific Weltanschauung began to take shape. With surprising rapidity the

---

4. *The Fourfold Vision* (London, 1945), p91; quoted by S.H. Nasr in *Man and Nature* (London: Allen & Unwin, 1976), p41.

new outlook crystallized in the minds of the pioneers and thence imposed itself upon society at large. It would seem that by the time of the Enlightenment, at any rate, Western man already found himself in a virtually desacralized and spiritually opaque cosmos. In place of a world 'full of lights manifesting the divine wisdom,' he now perceived 'the hurrying of material, endlessly, meaninglessly.' In the collective imagination the cosmos had become transformed—from a theophany—into that drab and problematic entity: the physical universe.

THERE IS A CELEBRATED TEXT in the Old Testament which from time immemorial has served as a mainstay of Judeo-Christian metaphysical reflection: it is verse 14 in the third chapter of Exodus. Let us recall the scene. Moses is tending his flock on the slopes of Horeb when suddenly, from the midst of a burning bush, he hears the voice of God. He reverentially approaches the site, and God speaks to him. And then Moses asks a remarkable question: he asks, in effect, what is the nature, characteristic or 'name' of God? And straightway the answer is given: *Eheieh asher eheieh*—Hebrew words, which the Vulgate renders as *ego sum qui sum*. Here is the full verse in English:

> And God said unto Moses, I AM WHO AM: and He said, Thus shalt thou say unto the children of Israel, I AM hath sent me unto you.

But what does this answer signify? What are we to make of it? Obviously it affirms that God exists. He exists, moreover, as a person, a unique 'I' in relation to all other beings; for He declares Himself to be a person, one who can say I AM. But there is more. There is an implication, and it is unmistakable: in truth, there is no 'other being': I alone AM.

This, undoubtedly, must be the crucial point: God alone IS. But how are we to understand this? 'It seems to me,' writes St Gregory of Nyssa, 'that at the time the great Moses was instructed in the theophany he came to know that none of those things which are apprehended by sense perception and contemplated by the understanding really subsist, but that the transcendent essence and cause of the universe, on which everything depends, alone subsists.'[5] But why? Does not the world exist? Are there not myriads of stars and galaxies and particles of dust, each existing in its own right? And yet we are told that the transcendent essence alone subsists. 'For even if the understanding looks upon any other existing things,' the great theologian goes on to say, 'reason observes in absolutely none of them the self-sufficiency by which they could exist without participating in true Being. On the other hand, that which is always the same, neither increasing nor diminishing, immutable to all change whether to better or to worse (for it is far removed from the inferior and it has no superior), standing in need of nothing else, alone desirable, participated in by all but not lessened by their participations—this is truly real Being.'

We are beginning to sense the metaphysical import of the Sinaitic teaching; and yet, who could say that he has grasped the point? We must not forget that 'the great Moses was instructed in the theophany' on the heights of Horeb, the *mountain of God* (Exod. 3:1); and as St Gregory observes elsewhere, 'the knowledge of God is a mountain steep indeed and difficult to climb—the majority of men scarcely reach its base.'[6]

What chiefly troubles us, it seems, is the concept of being, or of 'truly real Being', in Gregory's phrase. A Platonic notion, some will say; and this may well be true. But above all, it is a *nomen Dei*, the very 'name' that has been revealed to Moses. Now a name must have some connection—some affinity—with the object it designates. No wonder, therefore, that the concept of 'true Being'—one of the 'names of God'—should prove difficult, to the point of eluding our grasp. And indeed, the fact is that no philosopher, be he Christian or Greek, ancient or modern, has yet succeeded in explaining what Being is. 'What, then, can I do?' exclaims St Augustine. 'What that existence is, let Him tell, let Him declare it within; let the inner man hear, the mind apprehend that true existence....'[7] And another

---

5. *The Life of Moses* (New York: Paulist Press, 1978), p60.

6. Ibid., p93.

7. *In Joannis Evagelium*, xxxviii, 10; see *The Nicene and Post-Nicene Fathers* (Grand Rapids: Eerdmans, 1974), vol. vii.

Christian master writes: 'I have no doubt of this, that if the soul had the remotest notion of what Being means, she would never waver from it for an instant.'[8]

To BE SURE, we perceive the trace of Being in all that exists: and that is why we say, with reference to any particular thing, that it *is*. But yet that existence, or that contingent being, is not an absolute being: it is not the being that belongs to God alone. And why not? Perhaps the most eloquent answer is that the things of this world are mutable: they come upon the scene, we know not from whence; they grow, change and decay; and at last they disappear, to be seen no more. The physical cosmos itself, we are told, is a case in point: it, too, has made its appearance, perhaps some twenty billion years ago, and will eventually cease to exist. What is more, even now, at this very moment, all things are passing away. 'Dead is the man of yesterday,' wrote Plutarch, 'for he dies into the man of today: and the man of today is dying into the man of tomorrow.'[9] Indeed, 'to be in time' is a sure symptom of mortality. It is indicative, not of being, but of becoming, of ceaseless flux; for as Plato observed long ago, 'how can that which is never in the same state *be* anything?'[10]

This recognition implies that Being is unchangeable, and indeed, that Sameness, too, is a name of God. We do speak of sameness, of course, with reference to mundane existences, even as we speak of being. But here, also, the appellation proves to be unwarranted. Let us consider, for example, the so-called sameness or self-identity of physical existences. From the time of Newton—or if you will, of Leucippus and Democritus—it was conjectured that this presumed self-identity derives from the atomic constitution of matter. Now atoms, supposedly, are so small as to be indivisible, and being indivisible, were held to be constant and indestructible. In a word, they were thought to be the irreducible and permanent building blocks out of which physical things are compounded. These large-scale

8. *Meister Eckhart* (C. de B. Evans, trans., London: Watkins, 1924), vol. I, p 206.
9. *Moralia*, 329D.
10. *Cratylus*, 439E.

things, moreover, possess only a more or less transient and phenomenal reality, inasmuch as their atomic constitution, as well as their internal geometry, are constantly changing. What 'really exists', and what alone retains its self-identity, are the atoms. But as we have seen in Chapter I, that conception has ultimately proved to be erroneous. It turns out that neither the erstwhile atom, nor the fundamental particles into which it can be decomposed, have a true self-identity. In the words of Schrödinger,

> we have been compelled to dismiss the idea that such a particle is an individual entity which in principle retains its 'sameness' forever. Quite to the contrary, we are now obliged to assert that the ultimate constituents of matter have no 'sameness' at all.[11]

The point is obviously basic, and Schrödinger puts it in the most emphatic terms:

> And I beg to emphasize this, and I beg you to believe it: It is not a question of our being able to ascertain the identity in some instances and not being able to do so in others. It is beyond doubt that the question of `sameness', of identity, really and truly has no meaning.[12]

Indeed, identity *has no meaning* as a physical concept. For as we have said before, it is an incurably metaphysical notion, and in fact, a name of God.

This has been universally acknowledged, moreover, by the masters of traditional wisdom. In the words of St Gregory, 'that which is always the same, neither increasing nor diminishing, immutable to all change ... is truly real Being.' As concerns 'existing things', on the other hand, the teaching implies that these entities are always changing, always in a state of flux, so that their very existence is in a way a process of becoming, in which however nothing is actually produced. This has been said time and again, beginning with Heraclitus and the Buddhist philosophers. And there can be little doubt that it is true: even modern physics, as we can see, points to exactly

---

11. *Science and Humanism* (Cambridge: Cambridge University Press, 1951), p 17.
12. Ibid., p 18.

the same conclusion. Only there is another side to the coin which is not always recognized. Existent things—the very flux itself—presuppose what Gregory and the Platonists have termed 'a participation in Being'. The point is that relative or contingent existences cannot stand alone. They have not an independent existence, a being of their own. *In Him we live, and move, and have our being*, says St Paul, speaking to the philosophical Athenians. And St Augustine, musing upon the nature of created things, declares (addressing himself to their Author):

> I beheld these others beneath Thee, and saw that they neither altogether are, nor altogether are not. An existence they have, because they are from Thee; and yet no existence, because they are not what Thou art. For only that really is, that remains unchangeably.[13]

Indeed, the cosmos itself, in its totality, has not an existence independent of God. It is not another being, or a separate entity, standing apart from God and confronting Him, as it were. God alone Is: and that is the import of the Sinaitic revelation.

LET US OBSERVE, moreover, that yet another *nomen Dei* is clearly implied by the formula of Exod. 3:14: and that is Unity or Oneness. For He Who Is can only be one. Indeed, He must be *one-without-a-second*, as a Vedantic phrase has it. For He is one in Himself (as indicated by the singular pronoun 'I'), and 'without-a-second' by virtue of the fact that He alone Is.

Now the unity of God, no less than His being, is beyond human comprehension, inasmuch as that unity surpasses all the instances of oneness to be found in the world. We say, for example, that a nation has one ruler; and yet this ruler is one man among many. Or we speak of a composite thing as one whole; and yet this whole admits of numerous parts. But God is not one among many, nor does He admit of parts. No analogy, therefore, can realize the true oneness of God. But nonetheless, every instance of oneness does

---

13. *Confessions*, VII, 11.

exemplify, however inadequately, that absolute unity which is the prototype and source of all that we call unity or oneness within the order of creation.

And such relative or participated oneness is to be found everywhere. For oneness is indeed the inalienable concomitant of being, as the Scholastics have so often said: *ens et unum convertuntur*. Thus being and unity are inseparable; and this holds true, moreover, not only *in divinis*, but even with reference to mundane existences. And so, to affirm that a thing exists is to say that it is one thing; and if it be permissible to speak of degrees of existence, it could even be said that a thing exists to the extent that it is one. Thus an artifact, for example, exists in a higher degree than a cloud or a heap of stones, things which are somewhat ill-defined, and not sharply discernible as an individual entity; and again, it is evident that a living organism, by virtue of its stupendous unity, exists in a pre-eminent sense. Yet everywhere we encounter multiplicity along with unity, or more precisely, we encounter a multiplicity that partakes of unity in some measure. If the multiplicity did not partake of unity, moreover, we could in no wise encounter it, which is ultimately tantamount to saying that it could not exist at all. In a word, things both exist and are known by virtue of their unity. And yet multiplicity remains: it is not by any means cancelled by the manifested unity. Thus, for all its oneness, the living organism is yet composed of many members and countless cells, each of which exists by virtue of its own manifested unity. But beyond these partial and manifested unities there is an absolute and unmanifest unity from which they all derive, and to which each bears witness: and that is the supreme oneness of the One Who Is.

God is the ultimate cause, moreover, not only of all unity, but of every multiplicity as well. For multiplicity can in no wise exist apart from unity: it is the shadow of a partial or participated unity, one could say. And so, paradoxical as it may seem, 'what is supremely one is the universal principle of all multiplicity,' as St Bonaventure has observed.[14] This is not to say, however, that unity and multiplicity derive from what is supremely one in the same sense: for the

14. *The Soul's Journey into God* (New York: Paulist Press 1978), v, 7.

former derives therefrom by way of participation—or as an image from its prototype—whereas the latter arises, not through participation, but by default or incapacity. Thus it is always unity, and not multiplicity as such, that constitutes an *imago* Dei within the world: a reflection, however distant and fleeting, of His supreme and transcendent oneness.

This idea has been expressed, in one way or another, by every serious metaphysician. It is indeed 'the concept of unity' to which Plotinus alludes, the truth from which the materialists have 'drifted far'. The idea has been explained with unrivaled insight and eloquence by Dionysius, the renowned Christian author and authority in high matters, whose historical identity has been the subject of considerable debate in more recent times.[15] It will be enlightening to quote a characteristic passage from this ancient master, which speaks of oneness as an epithet of the supreme Godhead, and explains the cosmological significance of this particular *nomen Dei*:

> And the title 'One' implies that It is all things under the form of Unity through the transcendence of Its single Oneness, and is the cause of all things without departing from that Unity. For there is nothing in the world without a share in the One; and, just as all number participates in unity (and we speak of one couple, one dozen, one half, one third, or one tenth) even so everything and each part of everything participates in the One, and on the existence of the One all other existences are based, and the one cause of all things is not one of the many things in the world, but is before all unity and multiplicity and gives to all unity and multiplicity their definite bounds. For no

15. Author of *The Divine Names, The Mystical Theology,* and *The Celestial Hierarchies,* Dionysius had for long been identified with the Athenian of that name, who was converted to Christianity by St Paul on the Areopagus, as related in Acts 17:34. His writings (first mentioned in AD 533, at a council held in Constantinople) have exerted an enormous influence upon Christian theological thought. St Thomas Aquinas quotes Dionysius profusely, and Richard of St Victor refers to him as the foremost authority on the metaphysical interpretation of Scripture. In recent times his presumed identity has been disputed, and he has come to be referred to as the pseudo-Areopagite. Be that as it may, the fact remains that the author of the aforementioned treatises ranks among the undisputed masters of Christian wisdom.

multiplicity can exist except by some participation in the One: that which is many in its parts is one in its entirety; that which is many in its accidental qualities is one in its substance; that which is many in number or faculties is one in species; that which is many in its emanating activities is one in its originating essence. There is naught in the world without some participation in the One, the which in Its all-embracing Unity contains beforehand all things, and all things conjointly, combining even opposites under the form of oneness.[16]

Finally, we should not neglect to point out what in any case is implicit in all that has been said: the fact, namely, that the true goal or function of science is nothing but the discovery of unity in natural phenomena. An apple falls from a tree, and someone recognizes that this seemingly isolated event manifests a universal law. But what is a law of Nature, except a certain mode of unity? The object of science, therefore, is to reduce the multiplicity of phenomena to the unity of principles, and ideally—if that be possible—to the unity of one single principle. Certain recent developments, however, especially in the domain of physics, suggest that this ideal unity—to which we incline, as if by instinct, or by a 'categorical imperative' of the intellect—may not be realizable on a scientific plane. It appears that there are laws of complementarity—which as yet are only partially understood—that preclude the kind of unified theory which had been the dream of physicists since the days of Descartes. The fact is that science, for all its actual and all its potential accomplishments, must ever rest content with more or less fragmentary insights. The perfection of knowledge is simply beyond its ken. And this for the reason that the supreme unity, whose reflections we discern in all the laws of Nature, is itself beyond every law: for that unity belongs, not to the creation, but to God Himself.

CURIOUSLY ENOUGH, the universe exists, not simply by what it is, but also by what it is not; even as a sphere, for example, exists, not

16. *The Divine Names* (C. E. Rolt, trans., London: Society for Promoting Christian Knowledge, 1972), XIII, 2.

just by what it includes, but also by the immeasurable volume of space which it excludes. Not by accident, then, is the cosmos subject to bounds: for in the absence of bounds it could not exist at all. Like the geometric sphere, the things of this world exist precisely by virtue of that which restricts—or terminates—their existence.

To proceed further with this geometric analogy, let us observe how one goes about to determine a figure in the plane: a circle, for instance. To effect this construction, we need first of all to determine a point in the plane which is to be the center of our circle; and then we must construct a second point, so as to define a particular radius. Having accomplished this, we have determined a particular circle as the locus of points whose distance from the given center is equal to the length of the given radius. Prior to the construction, one may say, all was in a state of potency; there was neither a circle, nor even a single determinate point. In fact, the first determinate point— generally referred to as 'the origin', in mathematical parlance— sprang into existence quite abruptly, so to speak, through the construction itself: by its very first step. And clearly, that is a most remarkable step, if only one considers that there is nothing in the conception of the mathematical (or so-called Euclidean) plane which would in any way permit us to pick out, or to distinguish, such an element. The determination of the initial point is therefore an act which logically presupposes the geometrician, if one may put it that way. It is the geometrician himself, in other words, who imposes—as if by fiat—the basic determinations through which the figure in question is defined or constructed, beginning with the primary determination, or the so-called origin of space.

Now it has long been realized that geometric considerations of this sort are singularly suggestive, and admit in fact of a precise metaphysical transposition. For the cosmos—and all that it contains—is likewise determined by certain bounds, as we have said before; and this conception, moreover, entails three fundamental ideas: firstly, a principle of determination, or that which imposes bounds; secondly, a potential recipient of bounds, or that which is subject to limitations; and finally, the bound itself, as the determination that is imposed and received. The first, or active principle of cosmogenesis, we must understand, is none other than God, conceived as the Creator, Lawgiver or Architect of the world. It is He that creates, or determines, by His divine fiat, in accordance with the verse: *He spake, and it was done; He commanded, and it stood fast* (Ps. 33:9). The second, or passive principle, answers to the conception of matter: not, to be sure, in the contemporary scientific sense, but in the Scholastic sense of *materia prima*, which is pure potency, and not an existing thing. And lastly, the notion of bound answers broadly to the Aristotelian and Scholastic conception of form.

Going back to our preceding geometric considerations, it is now apparent that the plane as such corresponds to matter, or to pure potency; the constructed figure, to form; and the geometrician himself to the active, or creative, principle. These correspondences, moreover, are by no means adventitious, but spring from a profound and objective analogy between geometric construction and cosmogenesis, an analogy which, in turn, they bring to light. One should add that this analogy was well known to many of the ancient schools, and provides in fact an essential key to a correct understanding of traditional cosmological teaching. It is no doubt what Plato had in mind when he said, in the Timaeus, that 'God geometrizes always'; and again, it must be the reason why the famous motto 'Let no one ignorant of geometry enter here' had reputedly been inscribed over the portal of the Platonic Academy. Yet the underlying conception is not by any means peculiar to Plato, or to the Pythagorean legacy. It is to be found, as a matter of fact, in the major metaphysical traditions of mankind, beginning with some of the earliest Vedic texts. Thus the Rig Veda, for instance, declares in unmistakably geometric language—and long before Pythagoras— that 'with His ray He has measured heaven and earth.'[17] And let us not forget that the Old Testament, too, speaks of God in a similar vein, as in the celebrated *Dominus possedit me* passage in the Book of Proverbs, where it is written that He has *set His compass upon the face of the deep.* There is no sound reason to suppose, moreover, that all these striking concordances—which could be multiplied indefinitely—are due simply to historical influences or borrowings. The phenomenon, it would seem, can be perfectly well accounted

17. *Rig Veda*, VIII, 25, 18.

for by the universality of truth, and the innate objectivity of the human intellect.

RETURNING ONCE MORE to our preceding geometric construction, let us now observe that the resultant circle—the constructed circle, which can be 'swept out' by a compass—presupposes another: an ideal circle, namely, which serves as the model or prototype of the constructed figure. There is no getting around this fundamental duality: the particular presupposes the universal by force of logical necessity. It is true, of course, that the ideal circle does not exist in the same sense as the determinate figure. But it exists nonetheless—in its own manner—'within the mind or intellect of the geometrician,' as one might say. It is the model which he contemplates, so to speak, in the act of geometric construction; and so the construction externalizes, and at the same time particularizes, what already exists in another mode. There is a categorical difference between the two, and also a certain continuity; for the constructed figure, after all, exemplifies its archetype.

This brings us to the following question: under the metaphysical transposition—which identifies the mathematical plane with matter, the constructed figure with form, and the geometrician himself with the active principle of cosmogenesis—is there a metaphysical reality that corresponds to the ideal archetype of the geometric construction? Now one thing, at least, is evident from the start: this reality—if in any way it exists—must belong to the supra-formal order. And this implies that the geometric paradigm, if it corresponds analogically to anything at all, must signify a transcendent or acosmic reality.

Is there such a reality? Is there something in the nature of God, perhaps, which plays the role of an archetype *vis-à-vis* created forms? Does the geometric analogy actually carry that far? This is the great question. It is a problem, moreover, which needs to be faced: no metaphysical doctrine worthy of the name can side-step this issue. For in the final analysis, the intelligibility of the cosmos— and the very possibility of metaphysical thought—hinges upon this point.

Let it be said in passing that the great metaphysical traditions have not only addressed themselves to the question at hand, but have—without exception, we believe—answered in the affirmative. In one way or another, each has affirmed a transcendent metaphysical reality that reflects itself in created forms, and constitutes the essential content of forms—even as a constructed geometric figure reflects or manifests its archetype. Thus there can be little doubt, for instance, that this is what the so-called Platonic doctrine of Ideas was intended to express; only one should add that the issue has been hopelessly confused by rationalist protagonists and critics alike, who have failed to grasp that the doctrine is necessarily analogical. In other words, these interpreters have spuriously identified the Platonic Ideas with such things as the ideal circle, not realizing that these mathematical entities are in fact no more than images or analogues of the truly transcendent realities to which the authentic teaching alludes. It is simply the old fallacy of 'mistaking the finger for the moon,' to put it in the Chinese idiom.

BUT WHERE DOES Christian doctrine stand on the issue? Now, as we have said before, Christianity perceives the cosmos as a theophany: and this answers the question. For it affirms not only that there is indeed a transcendent paradigmatic reality, but that God Himself is the supreme Archetype, of which the cosmos—and all that it contains—is but a partial and imperfect likeness. All of Nature is 'but a glass' reflecting the Face of God.

To understand this a little more clearly, and to apprehend the scriptural basis of this teaching, let us reflect upon the celebrated Prologue of St John, which speaks of the Logos or Word of God. To begin with, it is to be noted that the divine Word corresponds by analogy, not to the outer word or words which are spoken audibly, but first and foremost to 'the word of the heart signified by the word of the voice,' as St Thomas explains.[18] Wherever anyone speaks with understanding, and wherever anyone hears with comprehension,

18. *Summa Theologiae*, I, 27, 1.

there must be that 'word of the heart'. Thus in speaking, the outer word is merely an expression of the inner, and in hearing, it is the external stimulus which causes the inaudible word to sound in the heart (it is thus 'the hammer that strikes my bell,' as Jacob Boehme has put it). In fact, both Christian and Oriental tradition distinguish various levels of the inner word, ranging from the outermost to the true word of the heart, which remains inseparable from the intellect itself. These are matters which can be understood through a kind of intellectual introspection, and one should add that the result is enlightening in the extreme. For as St Augustine tells us:

> Whoever can understand the word, not only before it is sounded, but also before thought has clothed it with imaginary sound, can already see some likeness of that Word of whom it is said: *In the beginning was the Word.*[19]

Now the creation as such is comparable to a word, for as Meister Eckhart observes, 'quite generally that which is brought forth by someone is his word: it declares, indicates and manifests that from which it proceeds.'[20] But it is likewise clear that the cosmos corresponds to the outer word—the word of the voice as opposed to the word of the heart—for the world has not the nature of God. Thus it exists, not in Being, but in becoming: all things are in a state of flux, as Heraclitus observed. Hence the world is not 'one in Being with the Father'; and so it is 'made, not begotten', even as the divine Word is 'begotten, not made'. Yet the cosmos is made in the likeness of the Word that was *in the beginning*, as with us too the spoken word is a certain likeness of the word conceived by the intellect. The creation is thus a theophany, wherein all things speak of God, 'for every creature is by its nature a kind of effigy of the eternal Wisdom,' as St Bonaventure declares.[21]

The act of creation, moreover, can also be understood by analogy to artistic production; for it will be noted that the artist produces the

---

19. *De Trinitate*, xv, 10.

20. *Expositio s. Evangelii sec. Iohannem*, i, 4; see *Meister Eckhart: Die deutschen and lateinischen Werke* (Stuttgart: Kohlhammer, 1936), vol. iii.

21. *The Soul's Journey into God*, ii, 12.

outer thing through an inward vision of the idea or exemplar, which pre-exists in him as 'the art in the artist', to use a Scholastic phrase. Thus the artifact is produced as a likeness or image of its exemplar, which as such remains in the artist, or better perhaps, in his art. As St Thomas expresses it, 'the knowledge of God is the cause of things. For the knowledge of God is to all creatures what the knowledge of the artificer is to things made by his art.'[22] One should add that 'the knowledge of God' is none other than the Word of which St John declares: *All things were made by Him; and without Him was not anything made.* Moreover, as the *Summa Theologiae* maintains:

> Since, then, the world was not made by chance, but by God acting by His intellect . . . there must exist in the divine mind a form or likeness of which the world was made. And in this the notion of an idea consists. [23]

Thus, as the existence of a thing derives from the absolute being of God, so also its 'whatness' or quiddity derives from a divine exemplar: 'hence it is that all things pre-exist in God, not only as regards what is common to all, but also as regards what distinguishes one thing from another.'[24] Yet it must not be supposed that the divine ideas or exemplars coexist in God as some vast multitude of separate entities; for as Aquinas explains, 'the divine essence is not called an idea insofar as it is that essence, but only insofar as it is the likeness or model of this or that thing. Hence ideas are said to be many, inasmuch as many models are understood through the selfsame essence.'[25] In other words, multiplicity pertains only to the things that are made and not to the exemplar, which according to St Thomas is none other than the divine essence itself. In terms of the geometric analogy to which we have referred in the preceding sections, one could say that the bound is one in itself but many in its participations: the 'measure' is one, but the measured things are many. Is this perhaps why the Rig Veda speaks of the divine 'ray' (by

22. *Summa Theologiae*, I, 14, 8.
23. Ibid., I, 15, 1.
24. Ibid., I, 14, 6.
25. Ibid., I, 15, 2.

which 'He has measured heaven and earth') in the singular form? Be that as it may, the point is made as clearly as one could wish in the *Summa*—under the question 'Whether the exemplary cause is anything other than God?'—where it is said, with reference to the ideas, that although they 'are multiplied by their relations to things, nevertheless, they are not really distinct from the divine essence, inasmuch as the likeness of that essence can be shared diversely by different things. In this manner, therefore, God Himself is the first exemplar of all things.'

But this entails that the bounds of Nature—the very measures in which the cosmos has been established—proclaim *the invisible things of Him*, as the Apostle affirms, *even His eternal power and Godhead.*

AMONG THE BOUNDS of Nature—or the measures by which the cosmos is brought into existence—none, certainly, is more fundamental than the temporal moment. Now we must understand, in the first place, that the moment is not a duration, however brief, but the bound of duration: for every duration is bounded by its beginning and its end. We must also realize, moreover, that duration as such has no existence apart from the things that endure, even as length does not exist apart from extended entities. Furthermore, every existing thing or process has a duration: for to exist in the world is to endure. And so it follows that the temporal moment constitutes a universal cosmic bound.

Here again the analogy to geometric construction has become evident. We see that temporal existence, too, is actualized through bounds: the very bounds, namely, which terminate that existence.

This basic and perennial insight has been obscured in our day by the Newtonian doctrine of time, a theory which hinges upon two misbegotten notions: firstly, the idea that time is 'an absolute homogeneous continuum'; and secondly, the belief that the moment is a part of time. Now homogeneous time is conceived as a kind of empty receptacle of events, even as space, in the Newtonian theory, is held to be the empty container of corporeal existences. And just as space is thought to be composed of an infinite number of points—a metaphysical misconception to which the 'analytic geom-etry' of Descartes had paved the way—so time is likewise envisaged as an infinite multitude of instantaneous 'nows'.

In a way it is surprising that an empirically oriented civilization which takes pride in its devotion to 'hard facts' should thus be com-mitted to a position that is utterly chimerical: for where, indeed, does one encounter this homogeneous time, not to speak of that infinite collection of temporal atoms? The moral, perhaps, is that everyone must have his metaphysics, even as everyone must have his religion too: our only real choice lies between truth and error. Let us add that the modern metaphysics of time does not, in fact, sit well in relation to twentieth-century physics. This science cries out, as it were, against these inbred notions, and yet they are 'stubbornly retained', as if they had been derived from an infallible authority. Thus our main intuitive difficulty with the pronouncements of rela-tivity theory stem precisely from our belief in an absolute and homogeneous time, made up of instantaneous 'nows'. That is the reason, after all, why we are startled to learn that 'absolute simulta-neity' has no physical meaning. For so long as we speak of actual durations there is no paradox, and no particular mystery. In fact, relativity theory can be viewed as a partial return, at least, from the notions of absolute space and absolute time, to space and time as actualized through a process of measurement. Slowly, it seems, but inevitably, modern physics is coming to realize that what exists is 'the measured'. And this, moreover, is in principle as far as science can ever go; for how could it grasp That which bestows the primary measure upon all things: the One, who with His 'ray' or 'compass', has established the cosmos?

Getting back to the subject of time, let us note that homogeneous time—like empty space—is a mere abstraction. Time as such is a potentiality, one can say: it is the potentiality, in fact, that is actual-ized by duration. And so the moment is not a part of time, but that which actualizes time: by dividing it, if you will, and thereby destroying its homogeneity. For as always, nothing that is perfectly undifferentiated—be it space, time, *prima materia*, or the once-debated ether—is able to exist.

The question arises now whether the temporal moment—as a

constitutive bound of all cosmic existence—has a theophanic significance. Does the moment of time, in other words, point beyond
itself to a transcendent paradigm? And if so, what is it that reveals
itself within the cosmos as that mysterious point which separates
the future from the past: as the 'now' that appears to move? It happens that the answer to this question has been known since ancient
times: the 'now' that seemingly moves is an image—a 'moving
image', as Plato says in the Timaeus—of the 'now' that stands still;
and that is eternity.

This perennial doctrine, admittedly, does not accord well with our
ordinary notions. But then the popular idea of eternity is itself
hopelessly confused, for it reduces evidently to the concept of 'endless
duration', which is an inherently contradictory notion, seeing that
duration is defined by its terminations. Now eternity is endless, to be
sure; but it is not a duration. Nor can we conceive of it as a limit by
envisaging a sequence of durations 'approaching infinity'. For it is not
duration—however long—but the instantaneous moment that
mirrors eternity.

What, then, is eternity? It is a state, or a plenitude of being, as
both St Augustine and Plotinus have observed, where 'has been' and
'will be' can find no place. There everything is concentrated within
a single point, as it were: it is being that fully owns itself, without
any scattering or dispersion. And yet it is not homogeneous, but
structured, if one may use that term; not empty, but perfectly full.

THE MOMENT, in its concrete actuality, manifests itself as the
present, the 'now'. But obvious as this may be if one only stops to
consider the matter, this fact is altogether lost in the abstract scientific way of looking at things. For in that perspective the moment
reduces in effect to a particular value of the time-coordinate, a value
which as such is in no way distinguished from any other. As the theoretician would put it, the equations of physics are invariant under
time-translations, and there the matter ends. But in reality it does
not: for it is clear that the present—by the very fact of being
present—is categorically differentiated from all other conceivable
moments. And this is neither a small nor a merely 'academic' dis-

tinction: the difference is as wide as that which separates what is from what is not.

It has often been said that all things exist in the present; and this is quite true: for what lies in the past has ceased to be, and what belongs to the future is yet unborn. Thus, curiously enough, it appears that the present—that seeming 'point' which has no magnitude at all—contains within itself all that ever was and ever shall be. One is strongly reminded here of the Gospel parable which likens the Kingdom of Heaven to a grain of mustard seed: for this 'least of all seeds' likewise contains within itself 'the greatest among herbs'. So, too, the present moment, which from a quantitative perspective appears as 'the least', is in reality 'the greatest', for it encompasses all that exists.

One may well object to this position on the grounds that cosmic reality is encompassed, not by a unique or single present, but by an infinitude of distinct moments, each of which assumes the status of presence, or of 'nowness', just once: for one single instant! But to think in these terms is to fall, once again, into the Newtonian fallacy: it is to conceive of the moment as 'a part of time'. We need not at this point enter upon a detailed critique of this thesis, a conception which has been the subject of philosophical debate since the days of antiquity. Suffice it to say that the Newtonian position can be refuted with compelling rigor via arguments which were known already to Aristotle—basically by showing that the concept in question leads to one of the 'paradoxes of infinity'.[26] And so the aforementioned objection—which presupposes that time is made up of moments—carries no force.

What the traditional doctrine maintains, as we have indicated before, is that the moment, so far from being 'in time' or 'a part of time', is that which actualizes time; it is not a duration, be it ever so

26. The idea of numerical infinity is inherently paradoxical. For example, if one were to speak of 'the number of integers' one would be forced to conclude (as Leibniz was perhaps the first to observe) that there are exactly as many even integers as there are even plus odd ones—a conclusion which by any count is absurd.

   The point is that infinite numbers—or if you prefer, infinite sets—do not exist. Now to be sure, modern mathematics does postulate such entities. It continues to

short, but the bound of duration. Thus the moment is not the tick of a clock any more than it is a day or a year. The bound is one thing, and the bounded entity another. And as always, multiplicity pertains to the things that are bounded—in this instance, to the durations—and not to the bound itself.

The world moves, while the 'now' stands still: this is the stupendous fact, so difficult to understand, and harder yet, by far, to realize in experiential terms. It appears that the cosmos is like a revolving wheel: the entire movement hinges upon the center, the one point which remains fixed. This is the ever-present 'now', the *nunc stans*, that wondrous pivot 'around which the primordial wheel revolves' (*punta dello stelo a cui la prima rota va dintorno*), in Dante's words. 'There every where and every when are focused ... Heaven, and all Nature, hang from that point.'[27]

STARTLING AS IT MAY SEEM, eternity is to be found—not in some distant and everlasting future—but in the ever-present 'now': like the Kingdom of God, it lies 'within'.[28] And so, too, eternity is to be

---

do so, moreover, despite the fact that the logic of this procedure has turned out to be far more precarious than was anticipated even a few decades ago, and that a growing contingent of mathematicians have abandoned these abstractions in favor of so-called constructive concepts. But even granting that it may be possible to operate with formal infinite sets in some logically consistent fashion, this does not in any way mitigate the absurdity of thinking that there are as many even integers as there are integers. The contradiction remains so long as the concept of number has not been formalized to the point where it no longer has any intuitive content at all. And when that happens we are no longer saying what we said before: we have then retreated into a purely conventional universe of discourse, an empty formalism which can at best be linked to reality through operational means. On the other hand, when the Newtonian doctrine speaks of an infinite number of moments, it is' not making a purely formal statement, much less is it setting forth an operational definition. What it asserts is quite evidently a metaphysical claim, which needs to be judged on its own ground: and that is precisely why the 'paradoxes of infinity' do come into play (even as they did two thousand years ago), and why the position is in fact logically untenable.

27. *Paradiso*, XIII, 11; XXIX, 12; and XXVIII, 41; resp.

28. One should add that this doctrine of time and eternity is not simply a private speculation or mere 'poetry'. As Ananda Coomaraswamy has demonstrated in

'entered' by way of the temporal moment, which is indeed 'the eye of a needle' through which it is hard to pass. It goes without saying that normally we are oblivious of this 'narrow gate'—this hidden dimension, one could say—which has to do with the way of the mystic, and with eschatology. We are habitually in the condition of the 'rich man'; like Martha, we are 'busy with many things.' We are strangers to that 'poverty of spirit' which the Gospel extols. And yet the gate is there, at the very center of our being. It is the 'heart' of which the mystic speaks, the 'closet' which Christ bids us enter (Matt. 6:6), the secret place where the saints commune with God. It is there, amidst the vicissitudes and dissipation of our life, even as it is there in moments of calm and recollection. It does not fluctuate, it does not move; unlike the things and creatures of this world, it is perfectly stable. For in truth it enshrines that immovable point, that 'pivot around which the primordial wheel revolves': the transcendent center 'where every where and every when are focused. . . .' Let us remember: 'Heaven, and all Nature, hang from that point.'

an extremely important study (*Time and Eternity*, Ascona: Artibus Asiae, 1947), it is integral to the Greek and Christian traditions, and may be discerned in Hindu, Buddhist, and Islamic teaching as well. Without doubt, it belongs to the perennial wisdom of mankind.

4

EVOLUTION:
FACT AND FANTASY

THE CENTRAL THESIS of Darwinism is the transformist hypothesis: the contention that one species can transform itself into another. How this may come about—through what causes or biological mechanisms—that is another question; the primary issue, in any case, is whether the higher species have evolved from primitive ancestors, and for that matter, whether a *bona fide* transformation of species has ever taken place.

Just as no two organisms belonging to a given species can be exactly alike, so too there is doubtless a certain variability—an elasticity, one could say—in the species itself. Thus it is certainly possible that a species may adapt itself to changes in the environment, or that it may develop certain beneficial traits. Now whether or not such transformations can in due time lead to the formation of a new species depends of course on what precisely one means by that term; and this is not a simple question. The issue has been debated extensively, and it is not clear whether there exists a single natural criterion (such as the ability to interbreed) which is in every respect satisfactory. But in any case it cannot be doubted that microevolutionary transformations do occur in Nature, whatever may be their extent as measured on the conventional taxonomic scale. The real question, thus, is not whether what we have defined to be a species is in fact invariable, but whether an evolutionary transformation can ever produce what we would unequivocally recognize as a new type of plant or animal. In other words, there is a grey area within which microevolution operates: what the transformist hypothesis affirms is that macroevolutionary transformations, too, have occurred.

As a scientific theory, the transformist claim is to be judged on the basis of observable facts. What, then, we must ask, are the principal sources of empirical evidence bearing upon this question, and what are the relevant findings?

In the first place one needs to consider the facts of paleontology; for it is evident that the fossil record constitutes indeed our one and only means of direct observation when it comes to the ancient forms of life. It is the telescope, so to speak, which renders the panorama of primordial life visible to some extent, and so provides a conceivable basis for the testing of evolutionist hypotheses. Here, etched in stone, are the hard facts with which the theory must accord.

Clearly, what the evolutionist would like to find in the paleontological record are chronologically ordered collections of fossils bearing all the earmarks of an evolutionary sequence: finely graduated chains, namely, exhibiting phylogenetic morphological variations as one proceeds from earlier to later specimens. Yet even though he should find such chains in abundance, he has still to establish their evolutionary origin; and it is obvious that paleontology itself can offer no warrant for this step. As the French biologist Louis Bounoure has observed, 'to see proof of a real descent in such a concordance between the placement of morphological types and their chronological position, is to adjoin to this concordance, which alone is the positive fact, the hypothesis of a filiation, whose verification is impossible and degree of certainty is always debatable.'[1] In other words, the transformist hypothesis is not directly verifiable in terms of paleontological findings.

On the other hand, it is likewise clear that a sufficient dearth of quasi-evolutionary fossil sequences would prove fatal to the theory. For if we suppose that the Earth has been populated for vast ages by plants and animals constituting transitional forms, and if it can be shown that during these periods the geological mechanism which accounts for the formation of fossils was operative, then it would stand to reason that the transitional forms should be represented in the paleontological record.

1. Louis Bounoure, *Déterminisme et finalité* (Paris: Flammarion, 1951), p 66.

But by and large they are not; and from the start, this has proved to be a major stumbling block for the protagonists of evolution. As the matter stood in 1859—and even more as it stands today—fossils do not make friendly witnesses for the evolutionist. Darwin himself, moreover, has perceived this very clearly. Thus, in *The Origin of Species*, he declares that 'this, perhaps, is the most obvious and serious objection which can be urged against the theory.' Repeatedly he raises the crucial question: 'Why then is not every geological formation and every stratum full of such intermediate links?' And his answer is this: 'The explanation lies, as I believe, in the extreme imperfection of the geological record.'[2] This, clearly, is the critical point—now as then—which the evolutionist needs to establish. 'He who rejects this view of the imperfection of the geological record,' writes Darwin, 'will rightly reject the whole theory.'[3]

One particularly troubling instance of the generic difficulty is the complete absence (or at the very least, the extreme paucity) of organic fossils in the Precambrian strata. Here is the problem, in Darwin's own words:

> There is another and allied difficulty, which is much more serious. I allude to the manner in which species belonging to several of the main divisions of the animal kingdom suddenly appear in the lowest known fossiliferous rocks. Most of the arguments which have convinced me that all the existing species of the same group are descended from a single progenitor, apply with equal force to the earliest known species. For instance, it cannot be doubted that all the Cambrian and Silurian trilobites are descended from some one crustacean, which must have lived long before the Cambrian age, and which probably differed greatly from any known animal.... Consequently, if the theory be true, it is indisputable that before the lowest Cambrian stratum was deposited, long periods elapsed, as long as, or probably far longer than, the whole interval from the Cambrian age to the present day; and that

2. Charles Darwin, *The Origin of Species* (Chicago: Britannica, 1952), p152.
3. *The Origin of Species*, p179.

during these vast periods the world swarmed with living creatures. . . . To the question why we do not find rich fossiliferous deposits belonging to these assumed earliest periods prior to the Cambrian system, I can give no satisfactory answer. . . . The case at present must remain inexplicable; and may be truly urged as a valid argument against the views here entertained.[4]

In the light of present geological knowledge one may add that the Precambrian strata amount to about four-fifths of the Earth's crust and correspond to a period of some 900 million years of geological history, beginning approximately 1,500 million years ago. Thus Darwin's surmise as to the enormous duration of the Precambrian age proves to be right: it is about one-and-a-half times as long as the entire interval from the Cambrian age to the present day. But this only accentuates the main problem. For so far as the fossil record is concerned, these gigantic Precambrian strata—which in some locations reveal over 5,000 feet of unbroken layers of sedimentary rock, ideally suited for the imprinting of fossils—have proved to be virtually blank. Admittedly, there have been sporadic reports of Precambrian finds purporting to derive from algae, bacteria or even 'worm-holes' caused by burrowing; but these have again been disputed, and in some instances, definitely disqualified. This contrasts with well over a thousand Cambrian genera, representing more than 5,000 species!

More plentiful than the Precambrian fossils, it would appear, are the theories which have been put forward to explain their absence. In a brief summary published in 1957 (which is undoubtedly far from complete), Dewar discusses no less than twelve theories of this type, all of relatively recent origin, and concludes that none of them is particularly cogent.[5] In any case, the very profusion of theories occasioned by the difficulty in question attests to the seriousness of the problem, and to the absence of any definitive solution.

---

4. *The Origin of Species*, pp 163–64.
5. See Douglas Dewar, *The Transformist Illusion* (Hillsdale, NY: Sophia Perennis, 2005), chap. 4.

THE BASIC DIFFICULTY—namely, a lack of intermediate forms—persists right through the fossil-rich Cambrian and later strata, as has been pointed out repeatedly, beginning with Darwin. The fact is that 'the majority of fundamental types in the animal kingdom present themselves to us without antecedent from a paleontological point of view,' as Deperet remarked in 1907;[6] and 'it remains true, as every paleontologist knows,' reiterates Simpson half a century later, 'that most new species, genera and families, and that nearly all categories above the level of families, appear in the record suddenly, and are not led up to by gradual, completely continuous transitional sequences.'[7]

Naturally, the evolutionist is obliged to account for this circumstance in a way that will safeguard his theory, and as we have noted in the Precambrian case, this need has given rise to a profusion of special theories. During the present century, moreover, the problem has been considerably complicated due to the fact that various facile solutions have been ruled out through notable advances in paleontology and related fields. In particular, it has become far more difficult to plead 'the extreme imperfection of the geological record.' According to an interesting study by Dewar and Levett-Yeats, for instance, first published in 1932, it turns out that a surprisingly high percentage of extant genera within two sample groups (i.e., mammals and mollusks) are represented in the fossil record.[8] In the case of land mammals, for example, the percentages range from 100 in the case of European genera to 56 for the Australian; and as might be expected, the figures are still better in the case of marine mammals. But even for volant genera (i.e., bats), where one would expect the smallest likelihood of fossilization, one finds that some 26% of the 215 extant genera appear in the record. Considering the fact that genera constitute a rather fine gradation on the taxonomic scale, these data accord ill with the tenet of extreme imperfection.

As the matter stands, the only avenue of escape from the negative

---

6. C. Deperet, *Les transformations du monde animal* (Paris: Flammarion, 1907).

7. G.G. Simpson, *The Major Features of Evolution* (New York: Columbia University, 1953), p360.

8. See *The Transformist Illusion*, chap. 2.

evidence of paleontology seems to lie in some feasible concept of cryptogenesis or 'hidden evolution', of which a number of variants have been proposed. One possible approach (and this applies especially to the higher stages of evolution, corresponding to the fossiliferous strata) is to postulate special phases of development during which the transformation of species takes place with such rapidity as to elude detection via the fossil record. In line with this general idea, one encounters such concepts as Severtzoff's 'aromorphosis', Schindewolf's 'explosive evolution', Zeuner's 'episodes of intense evolution' and Simpson's 'tachytely'. Somewhat different types of cryptogenesis have also been considered, such as de Beer's 'clandestine evolution'.[9] Yet all these theories suffer apparently from the same fundamental drawback, which is simply the lack of positive evidence. The best that can be hoped for in this domain, it would seem, is to avoid obvious conflict with known facts.[10]

The same observation applies to the various genealogical trees that have been postulated from time to time, beginning with Haeckel's famous specimen. So far as the finer branches are concerned, the claim has frequently been made that these can be certified by an actual fossil sequence. But quite apart from the logical problem already alluded to (the fact that no such sequence could possibly attest to an actual filiation), there are other difficulties here, which are frequently overlooked. For example, it has been demonstrated that starting from a given collection of fossils belonging to some group, it may be possible to extract a number of entirely different quasi-evolutionary sequences, depending on whether one

9. Ibid., chap. 9.

10. There exists a remarkable theory, expounded by Teilhard de Chardin before the Congress on Philosophy of Science, held in Paris in 1949, which deserves to be mentioned. This doctrine ('dont on ne peut qu'admirer la désarmante ingéniosité,' as Bounoure remarks) resolves the dilemma with astonishing directness by postulating 'the automatic suppression of origins'. According to de Chardin, the birth of a new phylum is accomplished in a short span of time through a small number of individuals, all of slight stature and rather fragile composition, which disappear without a trace, a circumstance which purportedly accounts for the seemingly sudden appearance of the new phylum. 'Sans doute,' Bounoure goes on to observe, 'faut-il être touche de la grâce évolutionniste pour trouver ce raisonnement convaincant.'

chooses the structure of the teeth, let us say, or of the paws, as the relevant factor. Inasmuch as these sequences are not compatible with any one postulated genealogy, we must conclude that some of them, at least, are artificial. But then, by what conceivable criterion can we distinguish between artificial and genuine sequences? As Bounoure points out in this connection:

> Our mind can well in the study of tertiary mammals, for example, establish certain comparisons and certain ideal relations between members of these groups; this is likewise the task *par excellence* of comparative anatomy. But in most cases one goes beyond the facts if one interprets these relations as denoting a real filiation, an actual descent. Abel is of the opinion that in the entire domain of the animal kingdom there are no more than five or six series of forms which are authentically evolutionary, that is to say, admitting the hypothesis of an actual descent through gradual transformation.[11]

When it comes to the main branches, on the other hand, the matter becomes still more tenuous. For it is here, especially, that the discontinuous aspect of the fossil record comes into full play, and where in place of a conceivable transitional sequence, however uncertain, we typically encounter a blank. How, then, can one presume to bridge these gaps? Considering the virtual impossibility of doing so with even a modicum of scientific rigor, it is not surprising that claims to this effect should have aroused considerable controversy, and that some authorities, at any rate, have remained skeptical. Thus Bounoure, for one, has this to say on the subject:

> It would be to underestimate the imagination of the experts to believe that in the face of the cryptic origin of the great phyla they should be lacking in resources. Haeckel had already pointed the way by inventing theoretical ancestral forms— the Protovertebrates, Protoselachians, Protoamniota and Protomammals, which have disappeared in the course of ages, or which in the advance of paleontology will some day be

---

11. *Déterminisme et finalité*, p 57.

discovered; Haeckel was never embarrassed in 'populating the ancient seas and continents with schemata' [Koken]. Now, one can remark that the phylogenetic trees of the zoologist proceed, in a specious manner, from the same gratuitous imagination: the leaves do indeed represent groups of real beings, but the trunk and the large branches are only an illusion or a subterfuge, insofar as they establish an inexistent continuity between groups; they are only an hypothesis introduced to support another hypothesis, and on the whole have no more value than a *petitio principii.*[12]

UNLIKE OTHER SCIENTIFIC THEORIES, which enable one to predict previously unknown facts, and which can therefore be tested in a more or less cogent manner, the doctrine of evolution has virtually no predictive content. Basically, one, argues in behalf of the evolutionist contention simply by adducing known facts which the theory purports to explain (frequently, as we have seen, with the aid of other hypotheses, which have been introduced specifically for this purpose). Now it is clear that such an argument derives whatever cogency it may have from an auxiliary premise to the effect that the given phenomena cannot be explained equally well on any other reasonable basis. But this obviously poses a fundamental problem: just how does one decide whether a conceivable alternative is reasonable? Is it reasonable, for example, to postulate some form of teleological causation? Or is it reasonable to view the matter in a metaphysical or theological perspective? In practice, to be admissible in the eyes of the scientific community, an alternative must evidently accord with the prevailing Weltanschauung. Once again, therefore, we find ourselves in a situation where hidden assumptions prevail, and where any doctrine 'which does not implicitly presuppose this point of view is assailed as unintelligible.'

But even if we agree to remain within the confines of the scientific outlook, the aforementioned auxiliary premise proves to be suspect; for when it comes to the biological sphere, especially, our

12. Ibid., p64.

knowledge is generally insufficient to rule out in advance all but a single scientific explanation of a given phenomenon. Consider, for instance, the following argument:

> The indirect evidence of evolution is based primarily on the significance of similarities found in different organisms, which are explicable only if they have derived the feature in question, structure or functions, from a common ancestor during descent with modifications, for the laws of probability insist that the fundamental similarities can be traced only to one single origin.[13]

But actually the laws of probability can do nothing of the sort. What we know is that any two organisms belonging to some given group exhibit a host of anatomical, physiological and other types of homologies. Now the author is saying, in effect, that the likelihood of finding so many similarities would be very small if it were simply a matter of chance. And this is unquestionably true; in fact, it follows logically from the very definition of probability. But to conclude that the given correlations cannot be due to chance is not to say by any means that they must be caused by a common origin. Obviously there are other conceivable possibilities. For example, it is quite conceivable that every organism within the given group must perforce exhibit all these common features simply because no

---

13. After pointing out that the theory of evolution had originally been based upon indirect evidence, the article goes on to note that 'Recently, however, direct evidence of evolution has been observed.' But oddly enough, we are not told what this 'direct evidence' consists of, or where we might find it. The nearest reference that we can find in the article occurs under heading (8), which begins with the following observation: 'It would require very special pleading to pretend that paleontology does not represent objective evidence for evolution, but more direct evidence is now also available, first from cytogenetics.' Moreover, the matter is disposed of in one sentence relating to the genetic code of three species of Drosophila, without so much as a cross-reference! Is this the purportedly 'direct evidence' for evolution to which the author had previously referred? In any case, this bit of evidence is obviously as indirect as all the rest. It simply amounts to the observation that the ordering of genes in three kinds of fruit flies could be ac-counted for on the supposition that the third is descended from the second, and the second from the first. See 'Evolution', *The New Encyclopedia Britannica*, 1981.

other 'blueprint' would work, or work as well. In other words, all things considered, the given homologies may be necessitated by natural requirements. Now whether this is actually the case is not in question here. We say that this is a logical possibility, a conceivable explanation of the given phenomena, which does not conflict in the slightest with the so-called laws of probability, or with any other known principles. And this is all that we need to say: for it proves conclusively that the fact of strong correlation does not in itself entail the hypothesis of common origin.

FROM THE START, the facts of embryology have provided one of the principal arguments in support of the transformist doctrine. Darwin himself had suggested that one might 'look at the embryo as a picture, more or less obscured, of the progenitor, either in its adult or larval state, of all the members of the same general class.'[14] And a few years later Haeckel formalized this idea in his famous biogenetic law, also known as the law of recapitulation. It affirms that the embryo, in its successive stages of development, recapitulates the phylogeny of its species; or in more pictorial terms, that it ascends that hypothetical tree of life to which we have already made reference. But while it appears that this theory has commended itself, for some time, at least, to a majority of biological authorities, there have all along been voices of dissent, and even some notable advocates of evolution have eventually rejected the biogenetic law. In 1909, Sedgewick, for example, propounded arguments against recapitulation which in his judgment disqualify the theory. Again, some embryologists (including de Beer, the proponent of 'clandestine evolution') came to the conclusion that the matter actually stands just the other way round: that phylogeny, namely, is based upon ontogeny. And in fact, de Beer and Swinton went so far as to say that 'in spite of the exposure of the theory of recapitulation, its effects continue to linger in nooks and crannies of zoology.'[15]

Be that as it may, it will be of interest to recall at least a few of the arguments that have been mustered against the biogenetic law. We

14. *The Origin of Species*, p 225.

shall base ourselves on a study by Dewar (himself a former student of Sedgewick at Cambridge).[16] (1) It is generally admitted that there is no such thing as recapitulation in the embryonic development of plants. 'This is inexplicable if recapitulation be a law of nature, and if, as transformists believe, plants and animals are descended from a common ancestor.' (2) 'Transformists believe that birds are derived from ancestors which possess teeth, but no traces of teeth are found in any of their embryos.' (3) 'The head of the human foetus progressively lessens in relative size as it develops, instead of becoming progressively bigger as the evolution theory requires.' (4) 'While the growing embryo shows all the supposed ancestral stages of the urinary system, it shows none of the presumed stages in the transition of the respiratory system from gills to lungs.' (5) According to one of the mainstays of evolutionist doctrine, the modern horse is descended from an ancestor having five toes. Yet the embryology of the horse exhibits no recapitulation of a five-toed ancestor.

In this connection Dewar points out that 'this does not prevent transformists from asserting that the presence of a tail in the human embryo from the 5th to the 8th week of its existence is the recapitulation of the stage of a long-tailed ancestor. This is supposed

15. G.R. De Beer and W.E. Swinton, in *Studies on Fossil Vertebrates*. ed. T.S. Westoll (London, Athlone Press, 1958), p3. 'One of these "nooks", it appears, is our *Britannica* article, which refers to recapitulation as a *fait accompli* established by Darwin.' (The embryologist von Baer, incidentally. whom the article cites in support of the biogenetic law, must not be confused with the embryologist De Beer, who is against it.)

16. See *The Transformist Illusion*, chap. 15. One might add that Dewar ranks high among the serious scientific authors in England and America who have opposed the Darwinian theory. Yet the work cited was until recently (see p65, n5) extremely difficult to obtain. 'Only too often,' writes a noted historian of science, 'the works of such authors have been deliberately neglected or suppressed. A case in point is the book by D. Dewar called *The Transformist Illusion*, Murfreesboro, 1957, which has assembled a vast amount of paleontological and biological evidence against evolution. The author who was an evolutionist in his youth wrote many monographs which exist in the libraries of comparative biology everywhere. But his last book, *The Transformist Illusion* had first to be published in Murfreesboro, Tennessee (!) and is not easy to find even in libraries that have all his earlier works. There is hardly any other field of science where such obscurantist practices are prevalent.' See Hossein Nasr, *Man and Nature* (London: Allen & Unwin, 1976), p140.

to be recapitulated, but not the five-toed stage of the horse ancestry.' And regarding the embryonic tail itself, he makes an interesting observation:

> It is important to bear in mind that at an early stage, i.e., before the second month in man, the human (and indeed every vertebrate) embryo exhibits a length of intestine behind the vent or anus. He who asserts that the human embryonic tail is a relic of a tailed ancestor, must, if he be logical, assert that the post-anal gut is a relic of an ancestor that went through life having such a strange organ. Writers who dilate upon the human embryonic tail are usually silent regarding the post-anal gut.

One final example, selected from the wealth of material which Dewar has brought to bear upon the question of recapitulation, may suffice to complete this brief review. It relates to the presumed 'fish stage' in the development of vertebrate embryos, which 'transformists unfailingly cite' as one of the most conclusive pieces of evidence. 'The truth is,' writes Dewar, 'that the so-called fish stage of the embryo must be passed through for the same reason that during construction a four-storied building must pass through a two-storied stage.' He elaborates on this point in an informative passage, which we will quote at length:

> The so-called fish heart and gill-arches have to be formed because the head region of the embryo from a very early stage onwards, requires a copious blood supply. This necessitates the early formation of a heart or pumping organ and a simple system of blood vessels. These have to be formed before there is time to develop the four-chambered heart necessary to the higher animal. To accomplish this, one or other of two devices must be adopted. Either a simple heart must be developed to function while another complicated heart is developing, or the simpler heart must be so constructed that it can become transformed into a four-chambered heart while it is operating as a heart. In this case the latter course is adopted, and by a most ingenious arrangement this simple heart while it is continu-

ously working is converted into a four-chambered heart. In some other organs, such as the kidney, the former is adopted.

ANOTHER CELEBRATED ARGUMENT in support of the transformist theory is based upon the so-called rudimentary or vestigial organs. These are structures found in living species which are apparently superfluous. 'Organs or parts in this strange condition,' writes Darwin, 'bearing the plain stamp of inutility, are extremely common, or even general, throughout nature. It would be impossible to name one of the higher animals in which some part or other is not in a rudimentary condition.'[17] And here again the transformist perceives evidence in support of his position. In fact, the case seems particularly clear-cut and convincing. 'What can be more curious than the presence of teeth in foetal whales, which when grown up have not a tooth in their heads; or the teeth, which never cut through the gums, in the upper jaws of unborn calves?'[18] The intended implication, of course, is that these curious facts admit the transformist hypothesis as their one and only explanation. But here, too, the case proves to be far more complex than Darwin had imagined, and with an increase of knowledge the picture has changed. As Vialleton has pointed out:

> Certain of these [supposed vestigial organs] deserve special examination because they play a part that escaped the notice of Darwin. When he cited as truly vestigial organs the germs of teeth in the foetus of whales devoid of teeth in the adult state, and those of the upper incisors of certain ruminants, the gums of which they never pierce, he forgot that these germs in mammals, where they are very large relatively to the parts enclosing them, play a very important role in the formation of the bones of the jaws, to which they furnish a *point d'appui* on which these mould themselves. Thus these germs have a function.[19]

17. *The Origin of Species*, p 225.
18. Ibid.
19. L. Vialleton, *L'Origine des êtres vivantes* (Paris: Pion, 1929), p 164.

And by way of corroboration, the eminent French anatomist goes on to observe that 'the disposition [of these foetal teeth], their form and their number, different from those of other Cetacea, show that in the whalebone whale, far from being merely the relics of an extinct ancestor, they have an individuality and a causality peculiar to them, since they are multiplied and adapted to the length of the jaw.' One might add that nonetheless the myth of the whalebone whale's teeth has survived, and to this day is often cited in authoritative treatises as a kind of gospel truth.

Generally speaking, the prime difficulty with 'useless' organs is that they may eventually prove to be useful. As in the case of foetal teeth, the presumed vestige may well have a hidden use, perhaps only at some particular stage of embryonic development, or may be somehow necessitated by that development. There is really no such thing as a 'plain stamp of inutility'. There was a time, for instance, not so long ago, when the function of the endocrine system was virtually unknown, and when such organs as the pituitary and pineal glands could with impunity be paraded as vestigial. But with the advance of scientific knowledge the list of such candidates has shrunk considerably, and as the matter stands today, their number is small. Moreover, such traditional 'show pieces' as the splint bones of the horse, the lateral toes of artiodactyls, the eyes of cave-dwelling animals or the wings of sightless insects have either been disqualified, or at least have come under serious suspicion.[20] Even the vermiform appendix in man has become controversial. For as one authority admits, 'in view of its rich blood supply it is almost certainly correct to regard it as a specialized and not a degenerate structure.'[21]

It is also interesting to note that whereas much has been made of the so-called vestigial organs, the subject of nascent organs is rarely brought up. Yet as Dewar has pointed out, the theory of evolution calls, not for vestigial organs, but for nascent ones: for rudimentary structures, that is, which are not yet of any use, but will become useful in their developed state. But it appears that no such organs have

20. See *The Transformist Illusion*, chap. 12.
21. W. E. Le Gros Clark. *Early Forerunners of Man* (Baltimore: W. Wood & Co., 1934), p 205.

been identified, be it in the fossil record, or in living species. 'So far as I am aware,' writes Dewar, 'no fossil exhibits a nascent organ: the earliest known fins are fully developed, so are the earliest legs and wings, whether of insect, bird, bat, or pterodactyl.' And with regard to living species, he remarks that 'if these species be really evolving, the majority of them ought to exhibit nascent structures in all states of completion, from unrecognizable excrescences to structures almost ready for use. Not a single one seems to exist!' [22]

IT HAS BEEN SUGGESTED that the degree of genealogical affinity between members of various species may be reflected in actual blood affinities. Now it is easy enough to establish relationships between different types of blood. For instance, if small amounts of blood from one animal are injected into another a reaction will generally ensue, resulting in the formation of an anti-serum. And when mixed with other blood, this anti-serum will cause a precipitation of blood protein, which can be measured (say on a percentage scale). Thus if one begins with animal X, the anti-X serum will cause varying amounts of precipitation, which may be taken as a measure of the degree of blood affinity with the blood type X. Anti-human serum, for instance, will cause 100% precipitation in man, 64% in gorillas, 42% in orangutan, 29% in baboons, 10% in oxen, 7% in deer, 2% in horses and 0% in kangaroos. The unresolved question, of course, is whether these figures have anything whatsoever to do with genealogical relationships. And yet protagonists of evolution have been sorely tempted to conclude from such data that among the given species, our nearest relative must be the gorilla, after which comes the orangutan, the baboon, the ox, and so on. There was a time, in fact, when blood-precipitation data were officially interpreted in this way. As one of the early authorities explained in 1909: 'We have in this not only a proof of the literal blood-relationship between man and apes, but the degree of relationship with the different main groups of apes can be determined beyond possibility of mistake.'[23]

22. *The Transformist Illusion*, p166.

More recent expositions, on the other hand, tend to be markedly less dogmatic on this point. The *Britannica* article, for example, from which the above data were taken, merely affirms that 'these figures serve as measures of chemical resemblance and affinity.' But what kind of 'affinity': chemical, or genealogical? The author does not say. And still, inasmuch as the results in question have been presented under the caption 'Evidence of Evolution', the implication is unmistakable. At the very least, this material has been used as a kind of bait.

It may be interesting to note that the early enthusiasm in this area was sparked by extensive blood-precipitation data (involving some 16,000 experiments) published by Nuttall in 1904. Now it seems that in the excitement caused by the discovery of blood affinities with apes, other aspects of the results have been roundly ignored. There are humans, for instance, more closely related to certain monkeys than to their fellow men, and there arc men 'as nearly related to carnivores, rodents and ungulates as to their own kind.'[24] According to some data, one of our nearest blood relations appears to be the whale! It is hardly surprising that many contemporary treatises on evolution have quietly abandoned the genealogical interpretation of blood-precipitation experiments.

ANOTHER TRADITIONAL SOURCE of evidence must not be omitted from this résumé: namely, the breeding and genetic experiments, which throw light on the degree of variability of living forms. Darwin himself had been greatly impressed by the fact that new varieties of a given species can be produced through selective breeding, and in a sense this observation forms the starting point of his theory. In other words, selective breeding was for Darwin the prime model of the evolutive process. What breeding does in miniature, Nature can accomplish on a grand scale through the mechanism of natural selection: that was the gist of his idea. And so *The Origin of Species*

23. C. Schwalbe, in *Darwin and Modern Science*, edited by A.C. Seward (Cambridge: Cambridge University Press, 1909), p129.
24. See *The Transformist Illusion*, chap. 13.

commences with a chapter entitled 'Variation under Domestication', and one finds that the entire subsequent argument pivots on the concept of variability which this body of observations is intended to exemplify. But in evaluating his claims, we must bear in mind—as much in fairness to Darwin as in the interest of truth—that in 1859 modern biology was still in its infancy. In the absence of any information concerning genes, Mendelian inheritance, mutations, the endocrine system and other vital factors bearing upon the variability of living forms, one was hardly in a position to hazard vast extrapolations from the observable facts. Looking at the matter in the light of present knowledge, let us now consider what the genetic facts are, and to what conclusions they point.

Certainly it is to be admitted, first of all, that domestic breeding does not lead beyond the limits of the species. Thus, after thousands of generations of breeding, a dog is still a dog, and despite considerable variation in size, proportions, coloring and so forth, each variety obviously bears the characteristic imprint of the basic form. Moreover, it is well-known that as one proceeds from wild stock to the higher breeds, the production of new varieties becomes progressively more difficult: the potential for new forms, it appears, is not unlimited. And the picture remains substantially the same when it comes to scientific breeding experiments, such as the famous studies involving the fruit fly, *Drosophila melanogaster*. Although these experiments—involving millions of specimens and thousands of generations have produced freaks in abundance, it seems that no new species has been formed. Even the application of x-rays, which increases the mutation rate by a factor of about 15,000, has not altered this fact. With other species, too, the result has been the same. Despite massive efforts expended over the better part of a century, no one has apparently succeeded in effecting a clear-cut transformation of the natural species. Stated in positive terms, this is overwhelming evidence in favor of the stability of living forms. As Caullery noted already half a century ago, when he proclaimed *la stabilité expérimentalement constatée des organismes actuels*:

> Contrary to what one could imagine for some fifty years, recent research has rather confirmed the idea of the existing stability

of animal and plant forms, and has relegated their variations either to purely individual phenomena without retention in the hereditary line, or to a limited diversification virtually contained within the type of each species.[25]

On theoretical ground, the discovery of genes and the Mendelian mechanism of inheritance had dealt a serious blow to the Darwinian concept of unlimited variability. Then came the discovery of mutations, raising hopes that these 'quantum jumps' might prove to supply the needed flexibility. Yet this too has turned out to be disappointing to the evolutionist. In the first place, it soon became well known that mutations are almost invariably detrimental. As a Nobel laureate has put it, 'Most mutations are bad. In fact good ones are so rare that we can consider them all as bad.'[26] The expectation, therefore, that the prime mechanism for evolutionary progress should be found in a process which invariably goes in the wrong direction appears dubious from the start. But not only the direction, but also the magnitude of the mutational variations, has proved to be disappointing. 'One knows today,' writes Bounoure, 'through the studies of geneticists, that mutation affects only relatively minor details, and never carries beyond the cadre of the species.'[27]

The question arises whether the picture has changed substantially since 1973 following the discovery of gene-splicing or 'recombinant DNA research', as this technique is officially called. To be sure, extravagant claims have been put forward in great numbers, and as so often happens where evolution is concerned, the dividing line between fact and fantasy has been obscured. Thus we are told time and again that the so-called genetic mechanism of evolution has at last been laid bare, and that one is now in a position to understand precisely how evolution operates—as if we knew to begin with that such a thing as macroevolutionary transformations have ever taken place! There are those, too, who would put the case more modestly; it is claimed, for example, that 'molecular genetics surely gives a

25. M. Caullery, *Le problem de l'évolution* (Paris: Payot, 1931), p401.
26. Muller, *Time*, Nov. 11, 1946; p38.
27. *Déterminisme et finalité*, p71.

much better defense of Darwinism than is offered by paleontology,' a statement which says very little indeed.[28] But even this little seems to be premature; for as Edward Wilson, the Harvard evolutionist, stated at a recent meeting of the American Academy of Arts and Sciences (on the subject of 'Darwinism: The expanding synthesis with molecular genetics'): 'within a few years we will begin to see some answers to evolutionary questions at the molecular level.' Perhaps; but meanwhile it is to be admitted that as things stand molecular genetics 'doesn't have much to say about speciation, about macro-evolution or about rates of evolution,' as Wilson goes on to observe. There seems to be a widespread expectation among the experts that some day it will. 'Ultimately,' conjectures Rudolf Raff, 'evolutionary mechanisms will probably be explained in terms of gene structure and rearrangements, but there is a very long way to go.' To which one might add that before macroevolution can be explained at all one must first establish that it exists; and here too 'there is a very long way to go.' In the meantime—like it or not—an inviolable constancy of species remains as the overriding experimental fact.

ENOUGH HAS BEEN SAID to show that the doctrine of evolution is not by any means the well-founded scientific theory which it is generally held to be. It is true that a host of facts has been brought forward in support of the Darwinist thesis; yet from the start not a few scientists and thinkers—including some of the most prominent exponents of evolution—have recognized the weakness of the empirical argument. Thus Darwin, for one, was rather cautious in the expression of his views. 'Darwin himself never claimed to provide proof of evolution or of the origin of species,' admits the Britannica article; 'what he did claim was that if evolution has occurred, a number of otherwise inexplicable facts are readily explained.' And Haeckel, the continental theoretician and renowned popularizer of evolution, went so far as to write (to a scientific friend) that 'one can imagine nothing more absurd, nothing which indicates more clearly a total lack of

---

28. Quoted by Roger Lewin in 'Molecules Come to Darwin's Aid,' *Science*, 216 (1982): 1092.

comprehension of our theory, than to demand that it be founded upon experimental evidence.' Now it is doubtful that the British empiricists could have concurred with their continental colleague on that score. Yet it is clear that on both sides the theory was propounded largely on *a priori* grounds, and that whatever may have been the prime factors motivating and driving the evolutionist movement, the balance was tipped in its favor, not by any clear-cut evidence, but by rational and ideological considerations of various kinds. Dampier, for instance, a staunch evolutionist himself, admits as much when he writes:

> Haeckel and other materialists, and in their train Teutonic philosophers and political theorists, joined to create that Darwinismus which made many of his followers more Darwinian than Darwin himself.... Men accepted natural selection as a proved and adequate cause of evolution and the origin of species. Darwinism ceased to be a tentative scientific theory and became a philosophy, almost a religion.[29]

Leaving aside certain interesting implications contained in these remarks, it is clear, in any case, that the theory of evolution was a timely doctrine, and that conditions favoring its reception had been prepared in advance by some of the major currents of European thought. As Hossein Nasr, another historian of science, observes:

> Rarely in fact has a theory connected with a particular science had such wide acceptance, perhaps because the theory of evolution itself, instead of being a scientific theory that became popularized, began as a general tendency that entered into the domain of biology. For this very reason it soon gained acceptance more as a dogma than as a useful scientific hypothesis.[30]

To be sure, the dogmatic or a priori character of the doctrine remains largely unrecognized, and even in scientific circles the belief is still widespread that evolution has been empirically verified beyond reasonable doubt. And yet, surprisingly, the contrary is also

---

29. *A History of Science* (Cambridge: Cambridge University Press, 1928), p 280.
30. *Man and Nature* (London: Allen & Unwin, 1976), p 124.

admitted often enough. As a noted French biologist has put it (after informing us that 'we have never been present even in a small way at one authentic phenomenon of evolution'):

> I firmly believe—because I see no means of doing otherwise—that mammals come from lizards, and lizards from fish; but when I declare and when I think such a thing, I try not to avoid seeing its indigestible enormity and I prefer to leave vague the origin of these scandalous metamorphoses rather than add to their improbability that of a ludicrous interpretation.[31]

What this scientist is telling us, in other words, is that despite the 'indigestible enormity' of the transformist claim and the fact that 'we have never been present even in a small way at *one* authentic phenomenon of evolution,' he accepts the doctrine on *a priori* grounds ('because I see no means of doing otherwise').

The quasi-official position, on the other hand, omits all reference to an indigestible enormity, and maintains simply that the doctrine has succeeded in explaining in a perfectly satisfactory manner a host of 'otherwise inexplicable' phenomena. But quite apart from the intrinsic difficulty of determining just when a given phenomenon is 'otherwise inexplicable'—a matter on which we have commented before—this contention suffers on yet another count. For so far from being readily able to explain a multitude of facts, the evolutionist is actually forced to stipulate countless *ad hoc* hypotheses to save his theory. We have already come upon quite a few examples illustrative of this point: the lack of Precambrian fossils and the general paucity of connecting links; the quandaries associated with recapitulation; the absence of nascent organs; blood-precipitation data 'teeming with absurdities' (Dewar); and *la stabilité expérimentalement con-statée des organismes actuels* as the final embarrassment. Now in every instance the evolutionist has been able to counter with some special hypothesis, and usually, in fact, with a sizable collection of such theories. Confronted with the observed stability of living forms,

---

31. Jean Rostand, *Le Figaro Littéraire*, April 20, 1957. Quoted in Titus Burck-hardt, 'Cosmology and Modern Science', in *The Sword of Gnosis*, ed. J. Needleman (Baltimore: Penguin, 1974), p143.

for example, one can say that the time-span or the number of generations in question is too small to permit the manifestation of evolutionary transformations, or that the given species has now arrived at a stage where such transformations can no longer occur. But although there is little evidence in support of such stipulations and no agreement among the experts as to which are right, one nonetheless believes that there must in any case be some legitimate explanation of the unfavorable facts that safeguards the theory. And here again the aprioristic nature of the doctrine manifests itself. Thus, when it comes to the endless multiplication of *ad hoc* hypotheses, the evolutionist does not perceive this as an instance of begging the question, simply because his main tenet is never really subject to question in the first place: evolution is in effect a *fait accompli*, and not to realize this is to evince 'a total lack of comprehension of our theory,' as Haeckel said long ago.

By the nature of the case, the doctrine of evolution cannot be established on empirical ground; and by the same token it is also in a sense 'unfalsifiable', as some contemporary philosophers of science have pointed out. This is both its strength and its weakness: 'its strength as dogma, and its weakness as scientific truth,' declares Bounoure.

IT IS SCARCELY AN ACCIDENT that Darwinism established itself at a time when the Newtonian Weltanschauung had attained the zenith of its influence. There is an evident connection between the two doctrines, inasmuch as under the Newtonian premises Darwinism, in some form, becomes virtually inescapable. In a universe answering to the conception of a closed mechanical system, the possibilities are greatly reduced. Moreover, if it be assumed—as it was from the start—that the Earth itself came into existence at some remote time,[32] one has hardly any other way left to explain the genesis of life and the origin of species except in transformist terms. Under such auspices, the 'means of doing otherwise' are not at hand.

---

32. The concept of cosmic evolution had already been enunciated by Descartes in his *Principia philosophiae*.

So far as the general climate of scientific belief is concerned, it would seem that this situation has not changed significantly since the initial triumph of Darwinism. On the other hand, it is to be noted that with the collapse of rigorous atomism and the associated Laplacian determinism, the notion of a clock-work universe has lost its scientific sanction. One knows now that even an actual clock-work is based upon merely statistical laws. The real world thus turns out to be much less rigidly constrained by our physical conceptions than one had previously imagined, a fact which holds true especially 'in the small'. In a very real sense, it appears that Nature is far more mysterious in its workings than the nineteenth century had been led to assume. The very advances of physics have brought to light hitherto unsuspected limits relating to the explanation of natural phenomena in terms of any conceivable physical mechanism. As a matter of fact, there is now strong reason to suspect that the 'ordinary laws of physics' do not apply to the highly structured forms of matter to be found within the nucleus of a living cell.[33] We are referring especially to those gigantic molecules within the chromosomes—the genes—which control the entire structure and functioning of the organism. Now from a physical point of view, what mainly differentiates these substances from inanimate forms of matter is their basic aperiodicity. They resemble thus an elaborate painting wherein every dab of color plays a special role, in contrast to inorganic matter, which might be compared to a large piece of wall-paper, wherein some simple pattern is repeated over and over again. Now inasmuch as the ordinary laws of physics—the laws that we normally test and use—are inherently statistical, their applicability in the case of solids depends upon periodicity. By analogy, they apply to the wallpaper as opposed to the painting. Hence, 'from all that we have learnt about the structure of living matter,' writes Schrödinger, 'we must be prepared to find it working in a manner that cannot be reduced to the ordinary laws of physics.'[34]

33. A very readable discussion of this topic may be found in Erwin Schrödinger, *What is Life?* (Cambridge: Cambridge University Press, 1967).
34. *What is Life?* p 81.

This is not to say that no laws—or no physical laws—are operative within the biosphere. Wherever there is life, there is order; and indeed, a degree of order which vastly exceeds anything to be found in the inorganic realm. The fundamental problem, in fact, with which every living organism must contend, is to maintain this tremendous order in the face of ambient disorder; and one might add that all the vital mechanisms have seemingly been instituted just for the accomplishment of this task. Moreover, the organismal order is distinguished, not only in degree—as measured in terms of 'negative entropy'—but in kind: it is what Schrödinger calls 'order from order', as opposed to 'order from disorder'. And undoubtedly, this difference is of the most far-reaching consequence. When it comes to the mysterious thing called life, even in its simplest manifestations, one is confronted by an entirely new picture.

An especially noteworthy feature of living organisms is what might be termed the primacy of the whole. Now a whole is something that exhibits a multiplicity of parts. The analytic mind has a tendency, moreover, to reduce the whole to its parts, or in other words, to conceive of the whole as a mere aggregate or sum of its constituents. It will be noted that this point of view is entirely characteristic of atomism, and of classical physics in general.[35] But with the advent of quantum theory, the picture began to change. 'Modern physics has taught us,' wrote Planck in 1929, 'that the nature of any system cannot be discovered by dividing it into its component parts and studying each part by itself, since such a method often implies the loss of important properties of the system.'[36] And as one moves from inorganic to living structures, this principle assumes a position of paramount importance. Thus, in the biological domain, one arrives in fact at the very antithesis of the mechanistic hypothesis: here it is not the whole that derives from the parts, but it is the parts, rather, that derive their existence (as parts) from the given whole.

35. To the extent that a physical system can be described in terms of differential equations, it has in effect been reduced to 'the sum of its infinitesimal parts'. Hence all of classical physics presupposes such a reduction. In the case of quantum mechanics, on the other hand, it is only the wave function (and not the system itself) that is subject to description in terms of a differential equation.

36. Max Planck, *The Philosophy of Modern Physics* (London: Norton, 1936), p33.

The organism is of course divisible into myriad components; but yet it is clearly one organism, exemplifying one basic form. We know, moreover, that this form is inscribed within the nucleus of every cell by way of the genetic code, and that from these centers it controls every aspect of life. One might say that the form itself is the center around which everything revolves, and from whence every organic structure is accorded its proper function.

Now the great problem is to account for the origin of this form, or if you will, of this stupendous order. The Darwinist answer, essentially, is that order springs from disorder, or that the greater order derives from the lesser. Leaving aside the vexing question as to how primitive organic forms could have sprung from inorganic substances—the painting from the wallpaper—Darwinism maintains that the transformation of species is accomplished basically through the process of reproduction. Thus it seeks to explain the origin of new organic forms through a biological mechanism whose natural function is just the reverse: the preservation, namely, of the given form from which also the mechanism in question derives its entire force and efficacy. For our part, we would find it difficult to conceive of a theory more directly at odds with what modern physics and biology have to teach.

THE MYSTERY of the living organism resides in its form. Every part and every process of the organism, its entire four-dimensional structure, springs from that form. But what is this form, this principle of order from which the creature derives its life? To answer that question in a Christian key, one needs but to recall the rudiments of metaphysical doctrine: the momentous claim that the creation is a theophany, and that 'every creature is by its nature a kind of effigy of the eternal Wisdom,' as St Bonaventure has declared. It follows, then, that what we have called the basic form of the organism can be nothing less than the manifestation of an eternal archetype subsisting in the Logos or Wisdom of God. In the final count, what shines through the form as the principle of order and the source of life is the Logos itself.

Yet granting that the form exemplifies an archetype, the question remains how the various species of plants and animals inhabiting our globe have actually been brought into existence. Were the species created by God at some particular point in time, perhaps within two or three 'twenty-four hour days', as some fundamentalists believe? Or does Christian doctrine admit of other interpretations, more palatable to the scientific mind? Is it possible, in particular, to reconcile the Christian position with the transformist hypothesis?

To answer these questions, let us understand in the first place that the act of creation is not to be conceived in temporal terms. We must not think that God created the universe at some time in the past, be it six thousand or twenty billion years ago. The point is that time applies to the creation and not to God. So too God acts, not in time, but 'in the beginning' (Gen. 1:1), a term which signifies 'the instantaneous and imperceptible moment of creation,' as St Basil explains.[37] This 'beginning', moreover, which is 'indivisible and immediate',[38] is none other than the *nunc stans*, the ever-present 'now' about which we have had so much to say in Chapter 3. As Meister Eckhart observes, 'God makes the world and all things in this present now.'[39]

There is in reality no conflict, then, between the position that the species have been created simultaneously—'all at once'—and the apparently contradictory view that they have been brought into existence successively, in a certain temporal sequence. In the first instance we are looking at the matter 'from the standpoint of eternity'—*sub specie aeternitatis*, as the Scholastics would say—and in the second we are looking at the same thing from a temporal perspective. Admittedly, the second point of view is the one which accords with our normal disposition. It is difficult for us to understand how 'time gone, a thousand years ago is now as present and as near to God as is this very instant.'[40] But then it is scarcely a cause for astonishment that we should find it hard to understand such things!

37. *Hexa.* 1,6, PG29, 16CD; quoted by M. Askoul in *St Vladimir's Theological Quarterly*, vol. 12 (1968), p63.
38. Ibid.
39. *Meister Eckhart*, C. de B. Evans, trans. (London: Watkins, 1924), vol. 1, p209.
40. Ibid.

The fact upon which Christianity insists is that all living creatures without exception have been created by God: *without Him nothing was made* (John 1:3). But in saying that all things have been created 'in the beginning', we must bear in mind that 'it was an everlasting beginning' [41]: *My Father works till now*, Christ says (John 5:17).

This leaves open the question whether God created the original progenitors of every kind or species in some special way—directly, as it were—or whether He creates invariably through a concatenation of secondary causes. Now this question pertains to the *modus operandi* of the creative act as envisaged from a cosmological point of view, a matter about which Scripture seems to have rather little to say. The Genesis account, in particular, merely suggests in very rough terms that the manifestation of terrestrial life has taken place progressively, through an ascending sequence of living forms culminating in man.[42] As has often been observed, moreover, there is nothing in Scripture which clearly rules out the transformist hypothesis. We cannot say with absolute certainty that the transformation of species is impossible, or that it has never taken place. Indeed, if it be true that *God is able of these stones to raise up children unto Abraham* (Matt. 3:9), then why not lizards from fish, and mammals from lizards?

But the question is: has He actually done so? And according to the mainstream of Christian tradition, He has not. There is in fact a consensus of belief among Patristic and Scholastic writers to the effect that the original progenitors of every natural species were not formed through the usual chain of secondary causes—they were not

---

41. The phrase comes to us from Jacob Boehme.

42. The claim has sometimes been made that the Biblical account of creation (and other relevant Scriptural passages) can be interpreted so as to accord in the main with modern scientific findings. Arthur Neuberg, for example, declares that 'one could almost expound the entire natural development (*Naturentwicklung*), the inorganic as well as the organic, the physical as well as the biological, within the framework of the Genesis account' (*Das Weltbild der Physik*, Göttingen: Vandenhoeck & Ruprecht, 1951; p161); and Karel Clays, in particular, has recently published an outstanding study along these lines, which also examines the Biblical teaching in relation to the paleontological record. See K. Clays, *Die Bibel bestätigt das Weltbild der Naturwissenschaft* (Stein am Rhein: Christiana, 1979).

born 'from seed'—but were brought into existence in a special way, answering more or less to the concept of direct creation.[43] Thus, according to this doctrine, living creatures can originate in two ways: through a primary or 'vertical' mode of generation, which does not involve seed as an intermediate cause; and through a secondary or 'horizontal' mode, that is to say, by means of a natural process. But at the same time we must not forget that the natural process, no less than the primary generation, derives its entire efficacy from the power of God.[44] Thus, in the last analysis, the distinction between the two modes pertains to the realm of appearances: it does not affect the ultimate cause, which is the same in both instances.

What mainly troubles us in regard to primary generation is that we cannot see it happening, nor can we imagine how such a thing could take place. Yet this is nothing more than the predicament in which we invariably find ourselves *vis-à-vis* realities which transcend the confines of our world. To understand a phenomenon in a natural or scientific way, we must trace it back to secondary causes. But this is just what cannot be done in the case of primary generation. Perhaps there are no secondary causes here—as seems to be the case in miracles[45]—or perhaps the causes are too subtle to fall within our reach.[46] In either case, we stand before a prodigy, a phenomenon that shatters the normal illusion of a closed and self-sufficient universe.

43. The chief authorities are SS. Ephrem, Basil, Gregory of Nyssa, Chrysostom, Ambrose, Augustine, Bonaventure, Albertus Magnus, and Thomas Aquinas. References and translations of source material may be found in E.C. Messenger, *Evolution and Theology* (New York: Macmillan, 1932).

44. 'The power of generation belongs to God,' says St Thomas (*Summa Theologiae*, I, 45, 5); and again, 'The power of the soul, which is in the semen through the Spirit enclosed therein, fashions the body' (ibid., III, 32, 11). Moreover, this accords with the direct teaching of Christ (see especially John 6:63 and Matt. 23:9).

45. Despite a scientist belief to the contrary, miracles do happen, and perhaps more frequently than one might think. The fact is that a great many such occurrences have been authenticated beyond reasonable doubt. The canonization proceedings of the Roman Catholic Church, for example, provide a wealth of data bearing upon this question. Like other 'anomalies' that have proved to be scientifically enlightening, miracles, too, have something to say about the workings of Nature.

46. This was in fact the opinion of St Augustine and others, who attributed primary generation to the agency of *rationes seminales*. A good discussion of this

There are first origins, then, and indeed there must be. Every chain of secondary causes, traced backwards, must eventually lead to the brink of a mystery: even physical cosmology, it seems, has at last come to this recognition. Likewise, so far as biological chains of descent are concerned, there must always be a 'missing link': the only question is whether there are many—one for each natural species—or whether the branches of the genealogical tree trace back to one common primordial ancestor, so that the mystery of creation appears to be concentrated, so to speak, at a single point. Now as we have just seen, traditional Christian thought has opted for the former of these alternatives by positing two basic modes of generation. It is interesting, moreover, that modern theories of evolution have likewise converged to the conception of a two-phase process (the so-called *Zweiphasenhypothese*), wherein microevolutionary phases alternate with 'creative bursts' through which fundamentally new forms are brought into existence. The main point of difference between the modern and the traditional doctrine lies of course in the interpretation of these explosive or discontinuous happenings. It is evident, moreover, that the creationist interpretation fits the paleontological facts far better than the transformist hypothesis, insofar as it obviates the vexing problem of missing links. Thus the creationist is absolved from the necessity of postulating such a thing as Pierre Teilhard's 'automatic suppression of origins', nor does he require any other *ad hoc* hypotheses to explain away difficulties. Furthermore, the traditional doctrine is very well able to account for the existence of biological homologies; for as Titus Burckhardt has observed, 'by its deepest significance the mutual reflection of types is an expression of the metaphysical continuity of existence, or of the unity of Being.'[47]

One could add that this suggestion may prove to be enlightening even from a scientific standpoint. In the domain of vertebrate embryology, for example, the phenomena which evolutionists have

---

rather abstruse subject may be found in Etienne Gilson, *The Philosophy of St Bonaventure* (Paterson, NJ: St Anthony Guild Press, 1965), chap. II).

47. 'Cosmology and Modern Science', in *The Sword of Gnosis* (Baltimore, Penguin, 1974), p146.

sought to explain through the hypothesis of recapitulation can now be viewed in a very different light. For if it be the case that man occupies a central position within the animal kingdom—a fact which can be understood from a metaphysical direction—then it need not surprise us if this centrality manifests even on an ontogenetic plane. This would mean that man can be viewed ontogenetically as the central trunk of a tree, whose branches represent stages in the ontogeny of other living forms. In a profound and distinctly non-Darwinist sense it may thus be true that the more primitive forms of life have actually descended from man. This, quite possibly, may be the great fact of which the evolutionist picture is but an inverted image.

Let us remark that a scientific theory consonant with this position has in fact been proposed. It was promulgated by Edgar Dacqué, a German paleontologist of note,[48] who had become persuaded that man represents the primordial form (*Urform*) from which the main types of the animal kingdom have sprung. Now as one might have expected, Dacqué's theory has been severely criticized in professional circles, even though it is by no means irrational or unscientific. As Carl Jung observes, the problem lies elsewhere:

> From the standpoint of epistemology it is just as admissible to derive animals from the human species, as man from animal species. But we know how ill Professor Dacqué fared in his academic career because of his sin against the spirit of the age, which will not let itself be trifled with. It is a religion, or—even more—a creed which has absolutely no connection with reason, but whose significance lies in the unpleasant fact that it is taken as the absolute measure of all truth and is supposed always to have common sense upon its side.[49]

IN SHORT, there *are* 'means of doing otherwise'; but they have been ruled out of court. Moreover, there is a traditional Christian

---

48. *Die Urgestalt d. Schöpfungsmythos neuerzählt* (Leipzig: Insel Verlag, 1943).
49. *Modern Man in Search of a Soul* (New York: Harcourt Brace, 1933), p175.

doctrine concerning the origin of living forms which accords both with reason and with the facts; the hitch is that it accords not with the modern bent of mind, 'the spirit of the age, which will not let itself be trifled with.'

5

THE EGO
AND THE BEAST

IT IS BUT A COMPARATIVELY SMALL STEP from Darwin to Freud. Given that the human species derives from sub-human ancestors, it follows that its mentality, too, has evolved out of a sub-human rudiment: the rational from the non-rational, the self-conscious from the instinctual. Now if that be the case, it is but natural to suppose that the bestial psyche still exists in us, concealed behind or beneath the conscious mentality, as a living vestige of the animal stage. And so we arrive essentially at the Freudian id, the psychic substratum which Freud takes to be 'the core of our being.'[1]

It is true that Freud has significantly curtailed this notion by stripping the id of all faculties relating to perception of the external world and response to external stimuli: the Freudian id as such is not in touch with the outer environment. It knows only its own somatic needs, 'tensions' which it seeks to eliminate through an appropriate discharge of energy. 'Instinctual cathexes seeking discharge,' as Freud puts it, 'that, in our view, is all there is in the id.'[2] It appears that 'the core of our being' is not especially well endowed, and there is not a lot to be said about it. Freud himself makes this perfectly clear:

> It is the dark, inaccessible part of our personality... we call it chaos, a cauldron full of seething excitations... it has no

1. *An Outline of Psychoanalysis*, cited hereafter as *AOP* (New York: Norton, 1949), p108.
2. *New Introductory Lectures on Psychoanalysis*, cited hereafter as *NILP* (New York: Norton, 1965), p74.

organization, produces no collective will, but only a striving to bring about the satisfaction of the instinctual needs subject to the observance of the pleasure principle.[3]

Considering that there can be no animal life without some measure of selection, adaptation and control, it is clear from this account that even in the lowest animals the id needs to be complemented by another psychic formation which can act as an intermediary between itself and the external environment. Now according to Freud, this second component of our psychic make-up derives from the first. 'Under the influence of the real external world which surrounds us,' we are told, 'one portion of the id has undergone a special development. From what was originally a cortical layer, provided with organs for receiving stimuli and with apparatus for protection against excessive stimulation, a special organization has arisen which henceforward acts as an intermediary between the id and the external world. This region of our mental life has been given the name of *ego*.'[4]

To perform its function as an intermediary, the ego must of course communicate with the id. To begin with, inasmuch as the ego has no energy of its own, it is obliged to obtain its power from the id; and having accomplished this somehow (often, it seems, with the aid of devious stratagems), it must then set about to guide the organism towards the fulfillment of its natural functions, a task which involves the exercise of certain controls over the instinctual propensities of the id. In this respect the ego may be likened to a rider controlling his horse. But as Freud points out, the relationship between the ego and the id corresponds in fact to a situation which is far from ideal: for it turns out in this case that the rider is eventually obliged to guide his mount towards a destination which has been chosen by none other than the horse. 'The ego,' Freud maintains, 'must on the whole carry out the id's intentions.'[5] And again: 'The power of the id expresses the true purpose of the individual

---

3. Ibid., pp 73–74.
4. *AOP*, p15.
5. *NILP*, p77.

organism's life.'[6] In a word, the ego is little more than a mask, 'a kind of façade,'[7] behind which stands the id.

BEFORE GOING ON to consider some of the other basic conceptions pertaining to the Freudian doctrine, it may be well to pause and reflect a moment on what has been said thus far regarding the ego and the id. To begin with, let us observe that the Freudian teaching—surprisingly enough—has something in common with the Christian anthropology. They agree in fact on a profound truth which we generally overlook, and which is crucial to any deeper understanding of man. One could put it this way: in his egocentric state man has forgotten who he is. In this condition he does not rightly know himself. He identifies with the ego, and in so doing fails to recognize that the ego as such is merely a phenomenon—an effect or an image, perhaps, of what we are. And what is that true nature, the veritable 'core of our being'? It is here, in answer to this basic question, that Christianity and Freud part company. For the Christian the core of our being is to be located in the soul, or in the highest part of the soul, which is itself an image—not of anything temporal or contingent, but of God Himself. And this is the reason why Clement of Alexandria and many a saint could say: 'If a man knows himself, he shall know God.' Now Freud's answer to the perennial question 'Who am I?' is quite different: for him, as we have seen, the quest leads not to an *imago Dei* but to a 'cauldron of seething excitations' or a chaos of 'instinctual cathexes seeking discharge.'

This is not to suggest that such things as seething excitations or instinctual cathexes do not exist. Certainly it is to be admitted, from a traditional no less than from a Freudian point of view, that our psychic constitution is complex and admits of various levels. The essential difference, however, between the traditional and the Freudian psychology, lies in the fact that the former envisions a hierarchic order which entails not only a 'below'—made up of subconscious psychic layers—but also an 'above' consisting of what

6. *AOP*, p19.
7. *Civilization and its Discontents* (New York: Norton, 1962), pp 12–13.

might be termed the spiritual degrees. To be sure, in our present state these higher levels of consciousness are also obscured, no less than the Freudian id. And so what is for us the unconscious is made up of the most disparate elements: it comprises the very antipodes of the psychic spectrum. The ego, then, with its narrow and shifting band of consciousness, occupies a middle ground: it is situated somewhere 'between Heaven and Hell', or between that in us which answers to these respective designations. Hence, symbolically speaking, it is possible in principle both to ascend or to descend from the plane of the ego. To ascend, moreover, is to draw closer to the actual core of our being: it is to attain a higher degree of self-knowledge. And so, too, it is by way of a 'descent'—a deviation from the archetypal nature and a certain lapse into oblivion—that we have arrived at the familiar level of psychic existence—the ego, which we normally take to be our self. What is more, this downward movement has not yet reached its ultimate term: there is still a 'below' left into which it is possible to slide. Having been created 'thoughtful and wise in the image of God,' as Gregory of Sinai points out, man enjoys nonetheless the option of making himself 'bestial, senseless and almost insane.'

Now a better description of the Freudian id could scarcely be conceived. One may suppose, moreover, that to men of spiritual discernment the existence of such a 'nether realm' in us will come as no surprise. Freud's main contribution, therefore, lies in the fact that he has elevated this particular element of our psychic make-up to the status of a first principle: he has made it 'the core of our being'. What appears on the traditional maps as the lowest fringe of psychic existence—a mere shadow of that supra-physical light which resides within us as an image of God—has become in Freud's eyes our very *soul*. On closer inspection the Freudian doctrine turns out to be an inversion of the Christian truth.

BUT LET US GO ON. After formulating his ideas concerning the ego and the id, Freud himself came to the realization that something had been left out of account. After all, the life of man is not concerned exclusively with biological necessities and the exigencies of survival.

It has also a higher aim, which finds expression especially in the spheres of art and religion, as well as in countless actions and reactions belonging to our daily life. There must be something, therefore, in our psychic apparatus which corresponds to the ideal aspects of human culture, a structure which engenders and supports the various modes of idealism. Now it is obvious, in the first place, that the id *per se* can do no such thing. The ego, moreover, having emerged from the id under the influence of external perceptions, as we have noted before, is primarily concerned with the task of 'representing the external world to the id,'[8] a function which is needed for the survival of the organism. Thus, by its genesis and its *raison d'être*, the ego is a realist: it is concerned more with external realities than with norms or ideals. 'To adopt a popular mode of speaking,' Freud observes, 'we might say that the ego stands for reason and good sense while the id stands for untamed passion.'[9] And although it is perhaps more amenable to 'the finer things of life' than the blatantly bestial id, the ego as such can be neither a moralist nor an artist, nor even a reputable citizen of a civilized society. To account for the so-called higher side of human life, therefore, a new psychic structure is required, a structure which by the very fact of its ideal propensities must be somehow set off from the ego. Now this is what Freud refers to as the 'super-ego', so designated because it acts as an observer and a judge of the ego, and because it prescribes the norms which the latter is supposed to follow and the ideal which it is called upon to emulate. For this reason it is also sometimes referred to as the 'ego ideal'.

So far the doctrine sounds promising enough. To understand, however, what Freud is driving at, we must follow him as he attempts to describe and explain the genesis of this new psychic entity. And this brings us to the celebrated Oedipus complex: the remarkable theory which asserts that during a certain stage of infancy the male child experiences a desire to murder his father and have sexual relations with his mother, while the girl, on the contrary, turns against her mother and wishes to have a child by her

8. *NILP*, p75.
9. Ibid., p76.

father. To make matters still more complicated, every human being, according to Freud, is inherently bisexual throughout his life, so that even in infancy homosexual tendencies enter the picture. Accordingly, it turns out that the child is burdened, in fact, by a 'double' or 'complete' Oedipus complex, made up of four perverse longings. In the 'normal' course of events—and after many anxieties, frustrations and traumatic experiences—the Oedipus complex is finally 'dissolved', at which point 'the four trends of which it consists will group themselves in such a way as to produce a father-identification and a mother-identification.'[10] This metamorphosis is supposed to take place at the age of five or thereabouts, and is thought to give rise to the third basic structure of our psychic make-up:

> The broad general outcome of the sexual phase dominated by the Oedipus complex may, therefore, be taken to be the forming of a precipitate in the ego, consisting of these two identifications in some way united with each other. This modification of the ego retains its special position; it confronts the other contents of the ego as an ego ideal or super-ego.[11]

The super-ego, then, represents a kind of internalization of the bi-polar parental image. As 'heir to the Oedipus complex', it is an expression of what Freud refers to as 'the most important libidinal vicissitudes of the id.'[12] These are just the impulses, as we have seen, which express themselves in the Oedipus phase as the tendencies towards incest and patricide. Within the structure of the super-ego which they themselves have helped to produce, these 'libidinal vicissitudes of the id' will—presumably—find more acceptable channels of self-expression. 'By setting up this ego ideal,' Freud goes on to explain, 'the ego has mastered the Oedipus complex and at the same time placed itself in subjection to the id.'

At the end of this circuitous tour, we thus discover that the super-ego, despite its frequently sanctimonious appearance, constitutes

---

10. *The Ego and the Id* (New York: Norton, 196)), p 24.

11. Ibid.

12. Ibid., p 26.

but another projection of the id. Like everything else within the human psyche, it is no more than a façade for the beast in us, the 'obscure id' which makes up 'the core of our being'.

NATURALLY the question arises how Freud has been able to ascertain the truth of these startling conclusions. How is one to verify, for instance, that the super-ego—the vehicle of all ideal thought—arises from the dissolution of the Oedipus complex? Or how can we be certain that there is really an Oedipus complex in the first place? Freud has a great deal to say about the sexual fantasies of infants: but how does he know these things? How did he find out that when a small girl catches her first glimpse of the male organ she is forthwith smitten with a 'castration anxiety', feels 'at a great disadvantage', and 'falls victim to phallus-envy, which leaves ineradicable traces on her development and character formation....'[13] Is this a fact, a datum from which scientific conclusions may be validly drawn? Or is it only a surmise, a conjecture which itself stands in need of being anchored to observable facts?

It would indeed be difficult to claim that such things as the so-called castration anxiety and phallus-envy in the child are scientifically observable. Freud himself points out in this connection that 'one has opportunities of seeing little girls and notices nothing of the sort.'[14] Yet he goes on to assure us that 'enough can be seen in children if one knows how to look.' But what does this mean? Wherein consists this superior mode of looking? Is it not rather a matter of seizing selectively upon various facets of infantile behavior and interpreting these in accordance with certain preconceived ideas? One is reminded of Freud's celebrated remark on the subject of nursing infants: 'as it sinks asleep at the breast, utterly satisfied, it bears a look of perfect content which will come back again later in life after the experience of sexual orgasm.' Yet it seems that Freud himself does not take too much stock in the possibilities resulting from such instances of 'trained observation'. Thus he asks us to con-

13. NILP, 125.
14. Ibid., p121.

sider 'how little of its sexual wishes a child can bring to precon-
scious expression or communicates at all,' and goes on to note that
'accordingly we are only within our rights if we study the residues
and consequences of this emotional world in retrospect, in people
in whom these processes of development had attained a specially
clear and even excessive degree of expansion.'[15] But to say this is to
assume, in the first place, that the infantile fantasies in question
exist, and secondly, that they continue to grow and develop right
into adult life, where in fact they can attain 'a specially clear and
even excessive degree of expansion.' But is this not a classic example
of begging the question? One cannot but agree with Andrew Salter
when he refers to this entire train of Freudian thought as 'an ever-
mounting crescendo of faulty reasoning.'[16]

To make matters still worse, epistemologically speaking, it turns
out that we are unable to gain our end through a study of normal
adults, for it is especially in the abnormal adult, in the neurotic
patient, that these elusive phenomena can be definitively observed.
'Pathology,' Freud declares, 'has always done us the service of making discernible by isolation and exaggeration conditions which
would remain concealed in a normal state.' But this, of course, is yet
another hypothesis, another assumption which needs to be made in
order to shore up the Freudian argument. As in the case of homo-
sexuality and incestuous appetites in infants, it must be supposed
that these 'conditions' remain concealed in normal individuals. But
even with this additional hypothesis in place, our difficulties are far
from over. In fact, one might say that they have just begun. For in
trying to draw scientific conclusions out of the welter of testimony
that can be elicited from neurotic patients, one is forced more than
ever to select and interpret, in short to hypothesize, for it is not to
be supposed that even under expert analysis the patient can simply
recollect such things as the presumed Oedipus phase of his development. 'You will recall,' Freud tells us in this connection, 'an interesting episode in the history of analytic research which caused me
many distressing hours. In the period in which the main interest

15. Ibid.
16. *The Case Against Psychoanalysis* (New York: Citadel Press, 1963), p18.

was directed to discovering infantile sexual traumas, almost all my women patients told me that they had been seduced by their father. I was driven to recognize in the end that these reports were untrue.... It was only later that I was able to recognize in this phantasy of being seduced by the father the expression of the typical Oedipus complex in women.'[17]

It is worthy of note that these incestuous fantasies were elicited 'during the period in which the main interest [i.e., Freud's main interest] was directed to discovering infantile sexual traumas.' One cannot but wonder to what extent these perverse imaginings may not have been somehow suggested in the course of the analysis, especially if one considers such matters as transference and other more or less occult processes associated with psychoanalysis. But this is something that we will touch upon later, when we come to consider the psychoanalytic procedure as such. For the moment, we wish only to emphasize what we have said before: namely, that even after one has made all the necessary assumptions which permit one to regard the fantasies of neurotic patients as a legitimate testing-ground for theories about infantile sexuality, one has hardly moved one step closer towards a scientific vindication of the Oedipus theory. It is no wonder, therefore, that this doctrine has been rejected by most contemporary schools of psychology, and that even among avowed followers of Sigmund Freud there is a pronounced tendency to read new and different meanings into the old formula, so as to arrive at something more acceptable.

The cogency of Freud's arguments has been questioned from the outset by scientists and philosophers, including the many who found themselves in sympathy with his doctrine. Ludwig Wittgenstein, for example, while commenting with approbation on what he termed 'the charm' of the Freudian theory, insisted nonetheless that this teaching lacks scientific status. This is all very interesting, he says in effect, but how is one to verify that it is true? And Robert Sears, of Harvard, in a painstaking report commissioned by the

17. *NILP,* p120.

Social Science Research Council, summarized these misgivings in the following terms:

> The experiments and observations examined in this report stand testimony that few investigators feel free to accept Freud's statements at face value. The reason lies in the same factor that makes psychoanalysis bad science—its method. Psychoanalysis relies upon techniques that do not admit of the repetition of observation, that have no self-evident or denotative validity, and that are tinctured to an unknown degree with the observer's own suggestions. These difficulties may not seriously interfere with therapy, but when the method is used for uncovering psychological facts that are required to have objective validity it simply fails.[18]

Freud for his part was prepared to defend himself by insisting that 'the teachings of psychoanalysis are based upon an incalculable number of observations and experiences, and no one who has not repeated those observations upon himself or upon others is in a position to arrive at an independent judgment of it.'[19] To repeat observations upon oneself, of course, means to be psychoanalyzed, and what Freud is telling us, in plain terms, is that only the psychoanalyzed or the psychoanalysts are entitled to judge the truth of his doctrine. It is needless to say that this momentous claim did not go over well with his critics, and that where there might have been doubt before as to the scientific validity of the Freudian claims, it now became clear that whatever else one can say for or against the psychoanalytic doctrine, it is certainly not a scientific theory.

Yet it would seem that this appraisal has been reached mainly by the discerning few. Within wider circles, and especially among bohemian artists, actors, and *littérateurs*, the subtle distinction between science and fiction was generally overlooked. 'The net result,' says a contemporary psychologist, 'was a public relations campaign that millions of dollars could not have duplicated. Once

18. *Survey of Objective Studies of Psychoanalytic Concepts*, Social Science Research Council, Bulletin 51, New York 1943, p133.

19. *AOP*, p9.

analysis became fashionable among the writers, it was a brief step before their more impressionable readers were fretting impatiently in the analysts' busy waiting rooms.'[20]

Freud himself always made it a point to stress the scientific character of his ideas. Science, according to Freud, constitutes the only legitimate road to knowledge, and this recognition, moreover, has been made by science itself. 'It asserts,' we are told, 'that there are no sources of knowledge of the universe other than the intellectual working-over of carefully scrutinized observations—in other words, what we call research—and alongside of it no knowledge derived from revelation, intuition or divination.'[21] We are not told, of course, by what steps 'the intellectual working-over of carefully scrutinized observations' has led to this remarkable insight; but in any case, this is one of the fundamental dogmas of the Freudian world-view.

Besides science, of which psychoanalysis is the completion, if not the apotheosis, Freud recognizes three other domains of human culture: art, philosophy and religion, the 'three powers which may dispute the basic position of science,' and of which 'religion alone is to be taken seriously as an enemy.'[22] As for the first, it 'is almost always harmless and beneficent; it does not seek to be anything but an illusion.' Philosophy, moreover, despite its ambitious pretensions, is at least harmless, inasmuch as it 'has no direct influence on the great mass of mankind; it is of interest to only a small number even of top-layer intellectuals and is scarcely intelligible to anyone else.' That leaves only religion as 'an immense power' and a serious threat to the scientific enlightenment of mankind.

This brings us to one of the great Freudian themes: 'the struggle of the scientific spirit against the religious Weltanschauung.' It seems to be a very serious matter for Freud, and as might be expected, he perceives that it was psychoanalysis which finally was able to gain the palm of victory for the side of science. 'The last contribution of the criticism of the religious Weltanschauung,' he declares, 'was effected

20. *The Case Against Psychoanalysis* (cited in n 116, above), p 11.
21. *NILP*, p 159.
22. Ibid., p 160.

by psychoanalysis, by showing how religion originated from the helplessness of children and by tracing its contents to the survival into maturity of the wishes and needs of childhood.'[23] In other words, the contents of all sacred belief, according to Freud, can be traced back to the Oedipus complex and its precipitate, the ego ideal. The latter, we are told, 'answers to everything that is expected of the higher nature of man. As a substitute for a longing for the father, it contains the germ from which all religions have evolved.'[24]

From this it might seem that Freud regards religion as one of the 'beneficent' illusions. Elsewhere, however, he expresses himself quite clearly on this question:

> Religion is an attempt to master the sensory world in which we are situated by means of the wishful world which we have developed within us as a result of biological and psychological necessities. But religion cannot achieve this. Its doctrines bear the imprint of the times in which they arose, the ignorant times of the childhood of humanity. Its consolations deserve no trust. Experience teaches us that the world is no nursery. The ethical demands on which religion seeks to lay stress need, rather, to be given another basis; for they are indispensable to human society and it is dangerous to link obedience to them with religious faith. If we attempt to assign the place of religion in the evolution of mankind, it appears not as a permanent acquisition but as a counterpart to the neurosis which individual civilized men have to go through in their passage from childhood to maturity.[25]

Questionable and unfounded as all this may be, it is said to carry the imprimatur of science. This is what awes us mortals. If psychoanalysis be science, how can the layman challenge its conclusions? Having once accepted this crucial dogma promulgated by Freud, one is prone to believe his other *ex cathedra* pronouncements as well.

23. Ibid., p167.
24. *The Ego and the Id*, p27.
25. *NILP*, p168.

MEANWHILE millions, it would seem, have done just that. Like the theory of evolution, Freudism too has entered into the mainstream of contemporary thought, and perhaps for the same basic reason: it constitutes a purportedly scientific doctrine answering to a major trend. This is not to suggest that Freud's teaching (which is after all incomparably more complex and hard-to-understand than the Darwinian thesis) has been accepted lock, stock and barrel by the masses. Nonetheless, numerous typically Freudian conceptions and attitudes have in fact found their way into the popular conscious-ness—the idea, for example, that culture is inherently 'repressive' and therefore bad, that morality is conventional and religious belief an illusion, and that at bottom the pleasure principle reigns supreme. These are the views, surely, which we imbibe in our schools and through the media. To think otherwise, moreover, is to run the risk of being categorized as a reactionary, a dullard, or quite possibly a neurotic.

It would be difficult indeed to overestimate the magnitude of the revolution spearheaded by Freud. It has undermined the remaining vestiges of Christian culture and succeeded brilliantly in its program of deconversion. As Philip Rieff observes, 'the systematic hunting down of all settled convictions represents the anti-cultural predicate upon which modern personality is being reorganized, not in the West only but, more slowly, in the non-West.'[26] It is beyond dispute, moreover, that Freud has contributed more to the establishment of this trend than any other single individual. 'Freud has systematized our unbelief,' writes Rieff; 'his is the most inspiring anti-creed yet offered a post-religious culture.'[27] One could say that a new type of human being has come into existence: 'psychological man'—the per-son who instinctively rejects all absolutes except the absolute of unbelief itself. And there are those who surmise that the new breed is destined to inherit the earth. 'Where family and nation once stood, or Church and Party,' Rieff predicts, 'there will be hospital and the-ater too, the normative institutions of the next culture. Trained to be in-capable of sustaining sectarian satisfactions, psychological man

26. *The Triumph of the Therapeutic* (New York: Harper, 1968), p 13.
27. Ibid., p 40.

cannot be susceptible to sectarian control. Religious man was born to be saved; psychological man is born to be pleased.'[28]

It is true that Freudian teaching, in the strict sense, has now been largely superseded. It is far too austere and too negative to sustain widespread popular support. The doctrine was exciting in the early decades of our century, at a time when the remains of Victorianism had not yet worn off. Today, on the other hand, it is 'selfism' in its countless forms—'the cult of self-worship' as Paul Vitz calls it—that has captured the popular scene. Not Freud but Fromm, Maslow, and Rollo May are the psychological gurus of the present day. And in certain respects their doctrine is very much opposed to the ortho-dox Freudian teaching which is not at all concerned with offering consolations. Nonetheless, it is clear that these later authorities are still following in the footsteps of the master, and that if it were not for the breach achieved by Freud, they could not have exerted any comparable influence upon society. Not until God and religion have been subtly dethroned in the popular imagination does the prospect of 'feeling good' appear quite so enticing.

Meanwhile the Freudian hostility towards religion, too, has become somewhat outdated. Once the therapeutic mentality becomes dominant within a culture it is no longer necessary to vituperate against Christianity, or against any other creed. One can then preach the gospel of 'pluralism' and 'toleration' with full confidence that in due time every facet of belief will be appropriately subjectivized and incorporated within a universal pantheon of therapeutic illusions. This fascinating possibility, moreover, seems not to have been entirely lost on the churchmen. Timidly at first, and then in droves, they have come forward to answer the call. As Rieff points out, 'the present ferment in the Roman Catholic Church' has indeed little to do with any renewal of spiritual perception, but constitutes 'a move toward more sophisticated accommodations with the negative communities of the therapeutics.'[29] In his view 'the sacralist yields to the analyst as the therapeutic functionary of modern culture.'[30]

28. Ibid., p24.
29. Ibid., p253.
30. Ibid., p77.

But let us get back to Freud and psychoanalysis. 'More often than not, contemporary psychoanalytic literature tends to brush aside the simple fact that psychoanalysis had its origin and sought its validation as a method of treating mental illness. Since many have tried, but nobody has yet convincingly demonstrated that psychoanalysis (or indeed any form of psychotherapy) is better for neurotic patients than doing nothing at all, this attitude is perhaps not surprising.'[31] What *is* surprising, however, is that this assessment should come (as it does) from a clinical psychiatrist. Yet Dr. Miller is by no means the first member of his profession to have reached that conclusion. Some thirty years earlier, for example, Abraham Myerson (a well-known practitioner) had this to say:

> I state definitely that as a therapeutic system, psychoanalysis has failed to prove its worth. First of all, it has not conquered the field as is the case with any other successful therapeutic approach as I have indicated in the first part of this paper. There is more reason to extol in the case of psychoses the pharmacological measures and the physiological stimulations than psychoanalysis. The neuroses are 'cured' by osteopathy, chiropractic, nux vomica and bromides, benzedrine sulfate, change of scene, a blow on the head, and psychoanalysis, which probably means that none of these has yet established its real worth in the matter, and surely that psychoanalysis is no specific. Moreover, since many neuroses are self-limited, anyone who spends two years with a patient gets credit for the operation of nature.[32]

One might add that Myerson had conducted a poll of neurologists, psychiatrists and psychologists to find out precisely what his colleagues thought about Sigmund Freud. The results disclosed a very broad spectrum of attitudes and beliefs, showing that expert opinion was more or less evenly divided concerning the worth—

31. 'Psychoanalysis: A Clinical Perspective', in *Freud: The Man, His World His Influence*, ed., Jonathan Miller (Boston: Little, Brown & Co., 1972), p112.

32. 'The Attitude of Neurologists, Psychiatrists and Psychologists Towards Psychoanalysis', *American Journal of Psychiatry*, vol. 96 (1939), p640.

both theoretical and therapeutic—of the Freudian teaching. It seems that just about every conceivable shade of opinion on the subject was represented in the response. There were those, for example, who praised Freud's theoretical insight but felt that psychoanalysis 'signally fails to produce beneficial results.' Again there were those who believed that 'the doctrine of infantile sexuality is completely against the facts,' and those who maintained that it can be objectively substantiated in large measure. There were those who believed staunchly that psychoanalysis is the panacea for most ills, and those who claimed that fewer than five percent of their patients could be benefited by Freudian methods. There were psychiatrists who maintained that 60 percent of the time psychoanalysis does more harm than good, and that four out of five analyses 'are not indicated.' There were those who hailed Sigmund Freud as the prophet of our age and those who regarded his pronouncements 'as one of the strangest anomalies and fantastic vagaries of the early twentieth century.' 'When one reads,' says Myerson, 'in Freud's *Civilization and Its Discontents* that woman has become the guardian of the hearth-fire because she is anatomically so constituted that she cannot put out the fire with a stream of urine, one wonders why there has been any acceptance of such doctrines.' One does indeed wonder! Meanwhile, whatever else may be gleaned from these sundry observations, such a phenomenal lack of agreement among the experts in itself suffices to prove that what we are dealing with is neither an authentic science nor a successful system of medicine.

While we are not aware of any more recent survey of a comparable nature, it appears that the prestige of Freudian psychoanalysis in professional circles has diminished considerably since Myerson's time. 'Except in France where Freud's concepts are still having important effects,' writes Vitz, 'the influence of psychoanalysis is declining. In the United States it has been under consistent criticism from almost all quarters for a number of years.'[33] The main thrust of this criticism is that the Freudian methods have on the whole proved ineffective in the treatment of mental disorders. 'In consequence,' says Rieff, 'there are new polemicists stalking Freud

---

33. *Psychology as Religion* (Grand Rapids, Eerdmans, 1977), p 13.

throughout the land. . . .'[34] One of the most vocal of these, we might add, is Thomas Szasz, another respected psychiatrist, who has gone so far as to argue that the very notion of 'mental illness' is erroneous and misleading. In line with this position, Szasz maintains that psychotherapy as such is not properly speaking a system of medicine, but a technique to influence and to control. He charges, moreover, that this technique is frequently put to immoral and harmful use, that this fact has been systematically obscured, and 'that all such interventions and proposals should therefore be regarded as evil until they are proven otherwise.'[35] Meanwhile, however, psychiatry is continuing to extend its sway both in the West and in the East.

IT IS INCONTESTABLE that psychoanalysis places the patient in a position of extreme vulnerability and subjects him to influences which he can neither understand nor control. As Salter has noted, 'the entire analytic procedure fosters the most complete and dangerous sort of dependence.'[36] It is well-known, moreover, that psychoanalysis owes whatever efficacy it may possess to the establishment of a special relationship between patient and analyst, a relationship known as 'transference'. As Freud has put it:

> The patient is not satisfied with regarding the analyst in the light of reality as a helper and adviser who, moreover, is remunerated for the trouble he takes . . . ; on the contrary, the patient sees in his analyst the return—the reincarnation—of some important figure out of his childhood or past, and consequently transfers on to him feelings and reactions that undoubtedly are applied to this model.[37]

In other words, the patient loses touch with reality and succumbs to a more or less infantile attitude, an attitude which gives the psychoanalyst power over his mind. 'If the patient puts the analyst in

---

34. *The Triumph of the Therapeutic* (cited in n 126).
35. *The Myth of Psychotherapy* (Garden City: Doubleday, 1978), p xxiii.
36. *The Case Against Psychoanalysis* (cited in n 116 above), p 145.
37. *AOP*, pp 65–66.

place of his father (or mother),' Freud explains, 'he is also giving him the power which his super-ego exercises over his ego, since his parents were, as we know, the origin of his super-ego. The new super-ego has an opportunity for a sort of after-education of the neurotic....'[38] In any case—whether we accept Freud's theories about the super-ego and its libidinal roots or not—the fact remains that by way of the transference the patient has opened himself to influences emanating from his analyst. To the extent, at least, to which these influences are consciously manipulated by the analyst, they are termed 'suggestions'. Thus transference paves the way to suggestion, and without a doubt this twofold process constitutes the central mechanism of psychoanalytic therapy. 'The influence of psychoanalytic therapy is essentially founded upon transference, i.e., upon suggestion,' says Freud.[39]

Let us pause to consider some of the implications of this startling admission. In the first place, it now appears that the patient's psychoanalytic testimony is likely to have been influenced by the analyst and his preconceived notions. As Freud himself explains: 'The mechanism of our curative method is indeed quite easy to understand; we give the patient the conscious idea (*bewusste Erwartungsvorstellung*) of what he may expect to find, and the similarity of this with the repressed unconscious one leads him to come upon the latter himself.'[40] But the existence of such a 'similarity' with repressed material is only an hypothesis, and a gratuitous one at that. What we know, and what alone is observable, is that the analyst makes suggestions, and that eventually the patient plays back the themes and images which had previously been implanted in his mind. Now this is easily explained, and without any further assumptions, through the fact that the patient, by virtue of the transference, has become pathologically vulnerable to the wishes and promptings of the analyst. He is virtually in the state of a hypnotic subject, prepared to act out whatever is suggested to him by the hypnotist. As

38. Ibid., p 67.

39. *A General Introduction to Psychoanalysis* (New York: Liveright Publishing, 1960), p 390.

40. *Collected Papers*, vol. 2 (New York: Basic Books, 1959), p 286.

Freud observes, a so-called positive transference 'alters the whole analytic situation and sidetracks the patient's rational aim of becoming well and free from his troubles. Instead of it there emerges the aim of pleasing the analyst, of winning his applause and his love.'[41] One is reminded in this connection of those unfortunate women who confessed to having been seduced by their father. Although one does not know how much applause and love they won in return for their incestuous fabrications, one can well believe that Freud himself was 'pleased'.

Apart from the fact that what has been said concerning the mechanism of transference and suggestion casts serious doubt on the objectivity of psychoanalytic findings, it also points out the terrible danger to which the patient exposes himself by entering willfully into the psychoanalytic pact. As is sometimes admitted within professional circles, even a perfectly normal person who subjects himself to psychoanalysis is bound to contract a *bona fide* neurosis as a direct consequence of the psychoanalytic process.[42] And it is needless to say that the more confused and wretched the patient becomes, the more susceptible he will be to the promptings of his analyst. 'With the harpoon of the transference in the patient,' says a clinical psychologist, 'the analyst can give him any interpretation, however preposterous, and the patient will usually go along with it.'[43]

But it appears that the acceptance of preposterous teachings may be the least of the dangers to which the hapless patient is exposed; to make matters worse, there is unquestionably an occult side to psychoanalysis. Even transference as such is something quite mysterious, something which one does not adequately understand. Freud seems to have sensed this at times, especially when in the course of his investigations he came upon certain strange phenomena. Thus, without committing himself on the point, he considered it likely that transference could bring into play hitherto unknown means of psychic communication and influence, such as telepathy.[44] But this

41. *AOP*, p 66
42. See *The Case Against Psychoanalysis* (cited in n16, above), pp 2–3.
43. Ibid., p 124.
44. See, for instance, *NILP*, pp 47–56.

means, in plain terms, that the psychoanalytic patient opens himself up to forces which even the analyst himself does not understand. It also means that somewhere along the line the analyst, too, may have become victimized by occult influences which are not under his conscious control. And this seems all the more likely if one recalls that, according to the Freudian tradition, the psychoanalyst is first of all to be analyzed himself.

Now what could be the nature and origin, let us ask in very general terms, of these mysterious forces which the scenario of psychoanalysis is designed to unleash? It would seem that one good look at the typical images which the process dredges up out of the unconscious should put us on the track. After all, Christianity has long proclaimed that there are indeed 'dark forces' within creation which can act upon our minds. 'What is this most hurtful whispering of the Enemy?' asks Tauler: 'It is every disorderly image or suggestion that starts up in thy mind.' Are we then to conclude that the Freudian id does in fact represent an infernal realm—that it constitutes a microcosmic exemplification, so to speak, of the nether world? As we have noted earlier, this does appear to be the case. And ironically enough, Freud himself has implied as much when he inscribed the following line from Virgil across the title page of his first major work: *Flectere si nequeo superos, acheronta movebo* ('If I cannot bend the gods, I will stir up hell').[45]

---

45. *Die Traumdeutung* (Vienna: Deuticke, 1900).

6

THE DEIFICATION
OF THE UNCONSCIOUS

THERE IS LITTLE DOUBT that Carl Gustav Jung ranks foremost among the students of Freud, as the master himself admits in a letter to his disciple wherein he speaks of 'anointing you as my successor and crown prince.' This investiture, as we know, did not take place, at least not in the way that Freud had intended; and it is also clear that by the time of their falling out, Jung had come to perceive his former mentor as one-sided, narrow and biased in his views. He felt, for instance, that Freud had greatly overplayed the role of sexuality and repression in the psychic life, and had exaggerated the importance of such things as fantasies and traumas experienced during infancy. Not that the premises of Freudian psychology were altogether unfounded: what mainly bothered Jung was the extreme dogmatism and exclusivism with which these concepts were upheld. He perceives Freud principally as an iconoclast, 'a great destroyer who breaks the fetters of the past,' an audacious deprecator of the nineteenth-century bourgeois milieu into which he was born, 'with its illusions, its hypocrisy, its half-truths, its faked, overwrought emotions, its sickly morality, its bogus, sapless religiosity, and its lamentable taste....'[1] But he does not perceive him as the prophet of a new age—a position which, as we shall see, Jung reserved for himself.

It is perhaps ironic that Freud, who took it upon himself to psychoanalyse the quick and the dead—from Moses to Woodrow

1. *The Collected Works*, Bollingen Series XX, cited hereafter as *CW*, (New York: Pantheon, 1971), vol. 15, p35.

Wilson—should in his turn be subjected to similar treatment at the hands of an apostate disciple. In any case, Jung makes it a point to explain Freud's idiosyncrasies as an over-reaction against the shams of a decadent civilization. Thus Jung sees the Victorian era as 'an age of repression, of a convulsive attempt to keep anaemic ideals artificially alive in a framework of bourgeois respectability by constant moralizings,' and believes that this explains, and to some extent justifies, the 'essentially reductive and negative attitude of Freud's towards accepted cultural values,' and more generally, his 'revolutionary passion for negative explanations.'[2] In particular, Jung thinks that the Victorian connection goes a long way towards explaining Freud's habit of harping incessantly on sexuality and on the sinister consequences of its repression. He accuses Freud of holding distorted views on the subject, and of knowing nothing but 'an over-emphasized sexuality piled up behind a dam....'[3] 'It is being caught in the old resentments against parents and relations,' Jung explains, 'and in the boring emotional tangles of the family situation which most often brings about the damming-up of the energies of life. And it is this stoppage which shows itself unfailingly in that kind of sexuality which is called 'infantile'. It is really not sexuality proper, but an unnatural discharge of tensions that belong to quite another province of life.'[4] In a word, Jung confirms what in any case one should have surmised: that Freud's views on sex are biased, provincial and rather sick.

In the light of these observations one is not surprised to find that Jung remains skeptical when it comes to Freud's scientific pretensions. He perceives the Freudian tenets more as an expression of subjective attitudes than as an objectively validated theory. What is more, Jung senses that somewhere along the road, whether consciously or unconsciously, Freud became diverted from 'serving science' to the accomplishment of 'a cultural task'. 'Today the voice of one crying in the wilderness,' Jung observes, 'must necessarily strike

2. Ibid., p34 and p35.

3. *Modern Man in Search of a Soul*, cited in n100 above, and hereafter as *MM*. (New York: Harcourt Brace, 1933), p121.

4. Ibid.

a scientific tone if the ear of the multitude is to be reached.... Secretly, psychoanalytic theory has no intention of passing as a strict scientific truth; it aims rather at influencing a wider public.'[5] Could it be that this is one of the initiatic secrets of psychoanalysis, and the true reason why no one who has not been initiated into the Freudian inner circle can rightly judge its claims? And was it perhaps as an erstwhile member of the Freudian brotherhood that Jung himself became privy to this fact?

Be that as it may, Jung is skeptical, too, with regard to the therapeutic efficacy of psychoanalysis. He considers the Freudian approach as being entirely too negative. 'Everything about it is oriented backwards,' he tells us. 'Freud's only interest is where things come from, never where they are going.'[6] Clearly, Jung does not share the Freudian faith that a regressive explanation of the patient's affliction is in itself sufficient to set things right. He freely admits that Freud has 'discovered all the filth of which human nature is capable,' but doubts that he was able to cure souls.

This brings us to another point of difference: the question of religion. Here, too, Jung accuses Freud of ignorance and bias. He charges that Freud knew nothing more than the 'bogus, sapless religiosity' of the Victorian era, with its 'sickly morality', that 'it is this sham religion that Freud has his eyes on.'[7] This is what Freud attacks with so much passion, and what he wishes at all cost to discredit as nothing but a bizarre manifestation of repressed sexual instincts. Jung, on the other hand, perceives the matter in a very different light:

> I do not doubt that the natural instincts or drives are forces of propulsion in human life, whether we call them sexuality or will to power; but I also do not doubt that these instincts come into collision with the spirit, for they are continually colliding with something, and why should not this something be called

5. *CW*, vol. 15, pp38–39.
6. Ibid., p37.
7. Ibid., p35.

spirit? ... As may be seen, I attribute a positive value to all religions.[8]

Whatever may be the ultimate nature of 'this something' that is called spirit, it is the crucial factor which enables us to transcend the recurrent exigencies of animal life and enter upon the fullness of human existence. 'If this is not achieved,' Jung warns, 'a vicious circle is set up, and this is in fact the menace which Freudian psychology appears to offer.'[9] The way of Freud does not lead beyond the tyranny of instinctual drives, 'this hopelessness', as Jung calls it. 'Wretched man that I am,' he exclaims, quoting the words of St Paul, 'who will deliver me from the body of this death?' And his answer to this perennial question is simple enough: 'There is nothing that can free us from this bond except that opposite urge of life, the spirit. It is not the children of the flesh, but the "children of God" who know freedom.'[10]

JUNG'S FINAL CRITICISM of Freud is that 'Freud has not penetrated into the deeper layer which is common to all men.'[11] Now that deeper layer is what Jung terms the collective unconscious: it constitutes our psychic inheritance, or that portion of it, at least, which is 'common to all men'. One should remark that Freud too speaks occasionally of an archaic heritage in precisely this sense, and likewise believes that such 'phylogenetic material' can manifest itself in dreams, myths and other cultural phenomena.[12] Thus, what Jung means when he accuses Freud of not having 'penetrated into that deeper layer' is not that Freud has failed to recognize the existence of a collective unconscious, but that he upheld superficial and fallacious views regarding its nature. His mistake, basically, was to picture the collective unconscious on the model of consciousness and its contents. But this is inadmissible, Jung maintains; for when it

8. *MM*, p119.
9. Ibid., p121.
10. Ibid., p122.
11. *CW*, vol. 15, p40.
12. See, for example, *AOP*, pp49–50.

comes to the collective unconscious we are confronted by something utterly foreign, something that is baffling, something incomprehensible.

These characteristics of the primordial psyche reveal themselves
most sharply in the case of insanity, which according to Jung is
nothing but a kind of forcible inundation of the conscious field by
the contents of the collective unconscious. Jung accuses his prede-
cessors of having been too much preoccupied with the study of
neurosis. If they had paid more attention to the phenomenology of
psychosis, he believes, 'they would surely have been struck by the
fact that the unconscious displays contents that are utterly different
from conscious ones, so strange, indeed, that nobody can understand them, neither the patient himself nor his doctors. The patient
is inundated by a flood of thoughts that are as strange to him as
they are to a normal person. That is why we call him "crazy": he
cannot understand his ideas. . . . The material of a neurosis is
understandable in human terms, but that of a psychosis is not.'[13]

The seeming irrationality of the deeper unconscious and its contents should not be interpreted in a pejorative sense. We must realize that it is the psychotic, and not the unconscious, that is insane.
The psychotic, moreover, is insane, not simply on account of the
ideas that have entered into his consciousness, but by virtue of the
fact that he is incapable of understanding these ideas. It is as if he
were confronted by a being of a different order—a god or a demon,
let us say, whose thoughts are not like our thoughts. Now this is just
what Freud had failed to grasp, and what invalidates his outlook
when it comes to the collective unconscious.

Like Freud, Jung too believes that the ego represents a
comparatively late formation, that it has evolved out of the obscure
depths of the unconscious through a gradual process of development
and dissociation. The birth of the ego, moreover, is also the birth
of consciousness, for 'consciousness needs a center, an ego to which
something appears.'[14] On the question whether the unconscious,
too, may have its center, Jung is decidedly skeptical. 'Everything

13. *CW*, vol. 9, pt. 1, pp 277–78.
14. Ibid., p 283.

points to the contrary,' he tells us;[15] it is precisely this absence of a center—of a 'personal consciousness'—he argues, that accounts for the fact that the unconscious presents itself as chaotic, irrational and incomprehensible.

Yet despite the profound difference between the conscious domain and the collective unconscious, there is an intimate contact between the two. Jung describes this interplay in the following terms:

> Normally the unconscious collaborates with the conscious without friction or disturbance, so that one is not even aware of its existence. But when an individual or a social group deviates too far from their instinctual foundations, they then experience the full impact of unconscious forces. The collaboration of the unconscious is intelligent and purposive, and even when it acts in opposition to consciousness its expression is still compensatory in an intelligent way, as if it were trying to restore the lost balance.[16]

As might be expected, the unconscious makes itself known to the conscious by way of images or ideas: it speaks to us, one might say, in a language of universal symbols. Jung is careful, moreover to distinguish between these symbols—which are objects of consciousness—and the unconscious contents which engender these conscious formations and express themselves by their means. It is this unconscious reality that stands behind the visible image or conscious idea which Jung calls an 'archetype'. The archetypes make up the content, so to speak, of the collective unconscious. 'They are living entities,' Jung explains,

> which cause the preformation of numinous ideas or dominant representations. . . . In reality they belong to the realm of the activities of the instinct and in that sense they represent inherited forms of psychic behavior. As such they are invested with

15. Ibid., p276.
16. Ibid., p282.

certain dynamic qualities which, psychologically speaking, are designated as 'autonomy' and 'numinosity'.[17]

The archetypes themselves, as we have said, are unknowable, inasmuch as they can never become objects of conscious experience. Nonetheless, they can be known indirectly through the images and 'numinous ideas' which they project. On this basis, moreover, Jung claims to have identified a number of specific archetypes: he has come up with quite a long list of them, in fact. Thus he speaks often of the *shadow*, the *anima* and the *animus*, three archetypes which occupy a particularly important place in his writings, or of the *wise old man, the great mother*, the *child*, and so forth, which are other archetypes. The point is that each archetype is supposed to have its own typical manifestations and its special function in the economy of psychic life.

It would take us too far afield to enter upon the particulars of this doctrine. Suffice it to say that the theory is assumed to have an explanatory value: basically, Jung operates with the archetypes much as Freud had done with his repression-born complexes. Thus, once again, all manner of psychic occurrences have become subject to interpretation based upon a specific algebra of psychological terms: a host of phenomena, relating to individuals as well as collectivities, can be accounted for, Jung believes, on the strength of the new psychological theory.

DESPITE RESERVATIONS *vis-à-vis* Darwinism, which Jung expresses occasionally, it is clear that he perceives the psyche in evolutionist terms. 'Just as the body has an anatomical prehistory of millions of years,' he writes, 'so also does the psychic system.'[18] And just as the consecutive stages of anatomical prehistory are recorded in the successive layers of the fossil record, so also there exists a record of our psychic prehistory, but with this notable difference: the earlier stages

17. *Psyche and Symbol*, cited hereafter as *P&S* (Garden City, NY: Doubleday, 1958), p xvi.
18. *Memories, Dreams, Reflections*, cited hereafter as *MDR* (New York: Pantheon, 1963), p 348.

of psychic life are with us still, not as dead fossils, but as the living contents of the collective unconscious.

It is interesting to recall that Jung came upon this conception by way of a dream, in which he found himself in a house of many levels. Descending to the basement, he discovered a hidden staircase which led down into a subterranean cave, filled with 'scattered bones and broken pottery, like remains of a primitive culture.' He related this dream to Freud, who was unable to interpret it to Jung's satisfaction. Finally, on the strength of his own interpretation, 'the dream became for me a kind of guiding image. , , , It was my first inkling of a collective *a priori* beneath the personal psyche.'[19]

Convinced of his discovery, Jung became more and more impressed by the magnitude of the psychic entity whose traces he was now eagerly investigating: a being 'transcending youth and age, birth and death, and, from having at his command a human experience of one or two million years, almost immortal.'[20] Jung was not slow to recognize that such a 'collective human being' would have super-human attributes, and might well be endowed with a potential knowledge and power of virtually godlike proportions. Moreover, if our individual consciousness has evolved—both in a phylogenetic and an ontogenetic sense—out of the collective unconscious, then this wondrous being is quite literally the parent of us all, and the giver of life. In a word, it began to dawn upon Jung that what he had come upon was nothing less than the numinous source from which all the religious conceptions of mankind have sprung, and to which they ultimately refer.

The facts relating to primitive religions seemed immediately to confirm this surmise. Thus it appears reasonable that archaic man, having but recently entered upon the ego-life, and being yet but little dissociated from the unconscious, should experience the numinous realm of the archetypes in very tangible and forceful terms. This explains, according to Jung, why the ancient forests and groves abounded with spirits, and why once upon a time the gods walked upon this earth. One may also suppose, moreover, that primitive

19. Ibid., pp 158–61.
20. *MM*, p186.

man felt threatened by these mythical beings, which after all represent the wild and chaotic forces from which he had just begun to emancipate himself, and that he consequently wished to appease these mighty spirits and secure their cooperation through sacrificial rites and magical practices, such as are to be found abundantly in primitive societies. And from here it might seem but a comparatively small step to the psychological elucidation of the higher religions, from Indian Yoga and Tibetan Buddhism to the sacred beliefs of Christianity.

But this was to be only a part of the ambitious program to which Jung felt himself called in the wake of his great discovery: besides interpreting the religious traditions of the past, he wanted also to understand in depth the crisis of the present age, and if possible, to discover a remedy. It had become clear to him that the progressive dissociation of the ego from the unconscious could represent no more than an initial phase of a larger evolutive process. He recognized, moreover, that this constitutes a hazardous step: for unless this trend is superseded in time by an integrative phase, it will lead eventually to neurosis and psychic disintegration. And in fact Jung became convinced that modern civilization had already entered into the red zone of collective neurosis: this, in his view, constitutes the root cause of our contemporary crisis. Basically, the problem is caused by a progressive alienation of the egocentric individual from the spiritual source of life: at bottom, our difficulty is religious in character. Almost all his patients above middle age, he tells us, suffer from a lack of purpose or meaning, caused by a lack of religious conviction and spiritual life. The ego has become imprisoned within its own narrow walls, and the springs of life are drying up. Jung believes, moreover, that Christianity, which in the past was able at least to offset these dangers on a collective scale, has essentially lost its meaning for modern man: it demands an act of faith which the person of today, indoctrinated as he is with scientific and humanist conceptions, is unable to achieve. The nineteenth century, while it had already suffered an erosion of Christian belief within the more educated strata of society, attempted nonetheless to maintain a Christian façade; and this gave rise to the deplorable sham against which both Nietzsche and Freud had reacted with such violence. But now the picture has changed: the twentieth century has openly given itself over to religious doubt, and what the sociologists term 'deconversion' is everywhere in progress. The result is that man has lost his spiritual guideposts: he has become disoriented and cut off from his own roots.

The time is now ripe, Jung believes, for a deeper understanding of the goal which Nature herself has set before us. This goal, he maintains, lies not in the glorification of the ego—in some ultimate victory over the dark forces of the unconscious, which is in any case an impossibility—nor does it lie in the annihilation of the ego, which would mark a return to unconsciousness. It lies rather in the harmonization of these two opposite or complementary aspects of the psyche, culminating in the birth of a single fully integrated organism. Jung maintains, furthermore, that this constitutes a perfectly realistic objective here and now, which can be effectively approached through the employment of appropriate means. The way to this goal is what he terms individuation: it is 'the process by which a person becomes a psychological "in-dividual", that is, a separate, indivisible unity or "whole".'[21]

And as might be expected, this is precisely what Jung's own system of psychotherapy is intended to promote.

We will not attempt to give a simple explanation of 'the process by which a person becomes an "in-dividual"': this is a rather complex and difficult subject to which Jung has devoted a great deal of space in his voluminous writings. Suffice it to say that the process involves what Jung terms 'an integration of the unconscious into consciousness,' and that this is to be achieved with the aid of archetypal images. Among these the circle and the square play a particularly important role: they form the basis of a symbolic diagram depicting the psyche in its totality. To the extent to which a person is able to grasp intuitively the psychological significance of such a 'mandala', he can attain to an effective realization of psychic integrality: a 'wholeness' which includes both the ego and the dark or hidden underside of the psyche. This realization, moreover, actualizes or

---

21. *CW*, vol. 9, pt. i, p275.

brings to birth a center which Jung refers to as 'the self'. Mysteriously, and not without travail, this psychic entity is born, and
becomes forthwith the goal toward which the process of individuation is consciously directed. The self has now become an interior sun
around which the ego circles, so to speak, and to which it is subordi-
nated. The illusion of egocentricity has thus been dispelled, and the
person has discovered 'the self': his own self, that is, 'what I really
am'. All that primitive man has worshiped in ignorance as an external pantheon of gods and spirits is now realized as an interior psychic reality: like the Kingdom of God, it is found 'within'.

It would be difficult to imagine how a doctrine of this nature could
be validated on a purely scientific basis, and as a matter of fact, Jung
himself desists from making this claim. Thus, while he speaks of
himself as an empirical psychologist, he is careful to point out that in
this domain, at least, empiricism involves a good deal of subjectivity,
and does not automatically safeguard against error. Indeed, it is one
of his criticisms of Freud that the latter has put forth his theories as a
kind of absolute and universal truth, oblivious of the special assumptions which underlie his outlook. 'At any rate,' Jung tells us, 'philosophical criticism has helped me to see that every psychology—my
own included—has the character of a subjective con-fession. . . .
Even when I deal with empirical data, I am necessarily speaking
about myself.'[22] And yet, despite this epistemological humility, it is
clear that Jung too has staked out claims, and gigantic claims at that.

It is in his posthumously published autobiography that Jung
gives us a closer look at the *modus operandi* of his psychological
research. In the form of intimate confessions he guides us through
a maze of enigmatic dreams and visionary apparitions, exhibiting
as it were the living world of psychic experience from which he
claims to have gleaned his main ideas. It all began with a series of
curious dreams which seemed to him to portend great truths, bearing especially upon the religious sphere. Later, after his split with
Freud, he resolved to enter upon a deliberate 'confrontation with

22. *MM*, p118.

the unconscious'; here is how Jung relates the beginning of this remarkable introspection, in which he was to be engaged over a period of some twenty years:

> It was during Advent of the year 1913—December 12, to be exact—that I resolved upon the decisive step. I was sitting at my desk once more, thinking over my fears. Then I let myself drop. Suddenly it was as though the ground literally gave way beneath my feet, and I plunged down into dark depths. I could not fend off a feeling of panic. But then, abruptly, at not too great a depth, I landed on my feet in a soft, sticky mass.[23]

Jung goes on to relate the strange spectacles which he beheld once his eyes 'grew accustomed to the gloom': there was 'a dwarf with a leathery skin, as if he were mummified,' a 'projecting rock', a 'red crystal', 'a stream and a floating corpse', 'a youth with blond hair and a wound in his head', and so forth. It appears that Jung immediately understood the import of all these revelations: 'I realized, of course, that it was a hero and solar myth, a drama of death and renewal, the rebirth symbolized by the Egyptian scarab.[24]

Such are the glimpses into his secret workshop with which Jung provides us posthumously. We are informed, moreover, that these dreams and visions had served from the start to reveal to him the substance of his psychological doctrines: 'the later details,' he tells us, 'are only supplements and clarifications of the material that burst forth from the unconscious, and at first swamped me.'[25]

This raises the question how it was possible for Jung to gain enlightenment from such an 'inundation' considering what he himself has told us *apropos* of psychosis. If the contents of the collective unconscious are 'not understandable in human terms,' and if to be inundated by such material is tantamount to insanity, then how did Jung himself escape this fate and emerge from his hazardous experiments not only sane, but enlightened? It appears that early in life Jung had come to realize that archetypal images *per se* are not

23. *MDR*, p179.
24. Ibid.
25. Ibid., p199.

enough: to ward off insanity and achieve enlightenment, one must have possession of certain keys which need to be obtained from a traditional source. Thus, soon after the dream that had set him on the spoor of the collective unconscious, he began with feverish interest to read through 'a mountain of mythological material, then through the Gnostic writers.'[26] He did not at that time, consciously at least, discover the keys for which he searched, but by his own admission, 'ended in total confusion.' In any case, we are told that later on, after he had made some progress in the interpretation of his visionary experiences, he felt a need to corroborate his conclusions. At this point he came upon alchemy: 'I had stumbled upon the historical counterpart of my psychology of the unconscious,' he writes. 'The possibility of a comparison with alchemy, and the uninterrupted intellectual chain back to Gnosticism, gave substance to my psychology. When I pored over these old texts everything fell into place: the fantasy-images, the empirical material I had gathered in my practice, and the conclusions I had drawn from it.'[27]

Jung seems to imply that the presumed concordances between his own conclusions and the Gnostic doctrines can serve somehow to validate both theories at one stroke. Thus he speaks of the necessity of finding 'evidence for the historical prefiguration of my inner experiences,' and adds that 'if I had not succeeded in finding such evidence, I would never have been able to substantiate my ideas.'[28] But it is not clear that his ideas have in fact been substantiated, with or without such 'prefigurations'. If it should be the case that others before him had reached similar conclusions, what does that prove? Is not truth more than just a matter of repetition? And if it happens that the Gnostics agree with Jung, what about all the other historical schools that do not? Finally, what assurances do we have that Jung was not influenced by Gnostic sources in the first place? He did study these writers assiduously before entering upon the development of his own theories, and even if these early perusals of Gnostic material had led to a state of 'total confusion', the encounter may

26. Ibid., p162.
27. Ibid., p205.
28. Ibid., p200.

nonetheless have left its mark upon his thought. In a word, when Jung claims to have substantiated his doctrine through historical prefigurations, this too is plausible only to the conditioned mind.

WHETHER IT BE A CASE of direct influence or of corroboration, the fact remains that Gnostic motifs play a leading role in the psychology of Jung. To begin with, Jung shares the Gnostic penchant for seeing all things in terms of so-called syzygies or 'pairs of opposites'—such as light and dark, male and female, good and evil, to mention but a few—as if cosmic existence itself were no more than a disturbed equilibrium, a process in which every plus must have its minus and every sum must add up to zero—if only we take care to include all the terms. Consonant with this outlook, the syzygies are said to arise from an undifferentiated state, which the Gnostics termed the Abyss (*bythos*) and which Jung for his part takes to be the collective unconscious. This is not to say that the two conceptions of the undifferentiated state are identical: we must remember that the Gnostics, in accordance with the objectivist tendency of ancient philosophy, thought of the *bythos* in objective or ontological terms, whereas the collective unconscious is naturally to be conceived in a psychological perspective. Yet the two notions are entirely analogous and play essentially the same role: thus the *bythos*, on the one hand, constitutes the ground of cosmic manifestation, whereas the collective unconscious represents the ground of psychological manifestation, and so of everything that can be introspectively observed. So, too, what the Gnostics view as a manifestation of cosmic existence or 'creation' in the Greek sense corresponds in the doctrine of Jung to an emergence into consciousness. In either case the genesis in question amounts to a differentiation into pairs of opposites of something inherently unknowable which resides in the ultimate ground.

Jung is very much concerned to apply these notions to the moral sphere. If everything must have its shadow side, and if existence itself results from a separation of opposites, then what we take to be evil can be no less essential than the good: like the two sides of a coin or the crest and trough of a wave, good and evil are but the complementary aspects of one and the same reality. It follows that the moral injunction 'to do good and eschew evil' becomes reduced to an impossibility: for in the total view of things the two sides of the scales are bound to cancel out. Furthermore, our endeavor to comply with the moral imperative can only serve to exacerbate the already existing disequilibrium, and must consequently lead to a crisis, an eventual breaking point. Thus it is clear that to accept the Gnostic axiom is implicitly to reject the Christian ethic.

Historically, of course, the opposition between the Gnostic and the Christian position is well-known. We must remember that the multi-faceted and somewhat polymorphic speculations subsumed under the heading of Gnosticism constitute one of the famous heresies against which Christianity has had to assert itself. In a way it was perhaps the most crass of all the heresies, the teaching that was most directly opposed to the central truth of Christianity. Thus Jung may be right when he views Gnosticism as 'the unconscious counter-position' to Christianity, and he may be right once more when he tells us that 'the spiritual currents of our time have, in fact, a deep affinity with Gnosticism.'[29]

Getting back to the question of good and evil, let us recall that in contrast to the Gnostic tenet, Christianity perceives evil as a *privatio boni*: a mere absence or 'privation' of the good, and thus something which has no essence of its own. Now it appears that this Christian doctrine has been a special source of irritation to Jung, a veritable thorn in his side. In any case he attacks it at every opportunity and even at the cost of considerable digression, engaging himself in what is quite obviously metaphysical speculation. 'The *privatio boni* argument,' he tells us at the conclusion of one of these disputations, 'remains a euphemistic *petitio principii* no matter whether evil is regarded as a lesser good or as an effect of the finiteness and limitedness of created things. The false conclusion necessarily follows from the premise '*Deus = Summum Bonus*', since it is unthinkable that the perfect good could ever have created evil.'[30]

---

29. *CW*, vol. 7, p77; and vol. 10, p83. See also my article 'Gnosticism Today', first published in *The Homiletic and Pastoral Review*, and republished in *Teilhardism and the New Religion* (Rockport, IL: TAN Books, 1988), pp 233–245.
   30. *P&S*, p49.

With Jung, on the other hand, as with the Gnostics, it was a set-tled conviction—a kind of gospel truth—that God is the author of evil. This tenet is already implicit in the Gnostic concept of cre-ation: the notion that the cosmos arises from a separation of oppo-sites. For on this assumption it would indeed follow that the power which is responsible for the manifestation of good is likewise responsible for all the evil existing in the world. 'In the final analy-sis,' Jung tells us, 'it is God who created the world and its sins, and who therefore became Christ in order to suffer the fate of human-ity.'[31] In other words, according to the 'theology' of Jung, Christ atones, not for the sins of man, but for the sins of His Father! And in fact, Jung perceives Christianity as a kind of drama re-enacting the 'tragic contradictoriness' of God, and so, too, of the universe which He creates or projects forth from Himself.

To Jung the 'Christ myth'—like every tale or symbol incorporat-ing archetypal contents—is both true and important: his complaint is simply that it has not been correctly understood. To unravel the true significance of the Christian symbolism, it seems that we need to come into possession of the Gnostic keys. Only then can we understand what everything means—down to the smallest details of the sacred liturgy!

Jung maintains that much of the blame for this customary incomprehension falls upon theology, which has foisted upon the faithful certain erroneous interpretations and ideas, such as the ignominious *privatio boni* and the related postulate '*Deus = Sum-mum Bonum*'. These false and euphemistic conceptions, he says in effect, have blinded us to the obvious truth that God is ambivalent, that He has also a dark side, and that He alone is responsible for the sufferings of the world. Thus, what theology terms Satan or Anti-christ, is in reality just 'the other face of God.'

The time has now come, Jung believes, when this forgotten and ostracized truth must once more be brought to light. Christianity, as it is commonly understood, is too literal a creed to be credible in the present age. With the advent of science and the 'miracles' of technol-ogy man has become less naive, less gullible. He still needs a living

31. *MDR*, p 216.

myth, however, and what is more, he has need of 'the Christian message', which Jung considers to be 'of central importance to Western man.'[32] Only this message 'needs to be seen in a new light, in accordance with the changes wrought by the contemporary spirit.'[33]

But it appears that this 'new light' is really quite ancient; it is in fact Gnostic. Demonstrably so: for if it be true that 'the spiritual currents of our time have a deep affinity with Gnosticism,' then to conform Christianity to the contemporary spirit is to conform it *ipso facto* to Gnostic ideas. For Jung this means above all to recognize God's 'dark face' and so in effect to deify Satan. As Philip Sherrard has observed, 'Jung regarded it as his task to redeem the Devil.'[34] The thrust of Jung's theological speculations, it seems, was to install Satan as the Fourth Hypostasis in a divine Quaternary.

BUT THEN how does Jung—avowedly an empirical psychologist— gain access to theological turf in the first place? In other words, how can psychological observation, even if it should attain visionary proportions, enlighten us about transcendental realities? Jung's answer is that what we term philosophical, religious, or metaphysical truth is nonetheless an object of thought, and as such it is a psychic phenomenon. He enunciates this position many times; for example, in his 'Psychological Commentary' on *The Tibetan Book of the Dead*: 'It is the psyche,' he tells us here, 'which, by the divine creative power inherent in it, makes the metaphysical assertion; it posits the distinction between metaphysical entities. Not only is it the condition of all metaphysical reality, it *is* that reality.'[35]

It appears, however, that Jung himself is not entirely pleased with this radical conclusion. 'I do not mean to imply that only the psyche exists,' he says elsewhere. 'It is merely that, so far as perception and cognition are concerned, we cannot see beyond the psyche.... All comprehension and all that is comprehended is in itself psychic,

32. Ibid., p210.
33. Ibid.
34. *Studies in Comparative Religion*, 3 (1969): p37.
35. *P&S*, p286.

and to that extent we are hopelessly cooped up in an exclusively psychic world.'[36]

But although Jung has now retreated from the pan-psychism of his earlier statement by admitting the existence of a non-psychic or trans-psychic reality, he remains nonetheless caught up in the fundamental contradiction of an implicit bifurcationism. Thus, on the one hand, he affirms that we are 'hopelessly cooped up in an exclusively psychic world,' while on the other he obviously believes in the existence of a physical universe and seems to accept what science has to say about it. At times, moreover, he goes so far as to suggest that

> the deeper 'layers' of the psyche [become] extinguished in the body's materiality, i.e., in chemical substances. The body's carbon is simply carbon. Hence 'at bottom' the psyche is simply 'world'.[37]

But this too seems not to be the last word. Elsewhere, for example, when he castigates 'the irresistible tendency to account for everything in physical terms,' he appears once again to be rejecting the materialist stance:

> Today the psyche does not build itself a body, but on the contrary, matter, by chemical action, produces the psyche. This reversal of outlook would be ludicrous if it were not one of the outstanding features of the spirit of the age. It is the popular way of thinking, and therefore it is decent, reasonable, scientific and normal. Mind must be thought of as an epiphenomenon of matter.... To grant the substantiality of the soul or psyche is repugnant to the spirit of the age, for to do so would be heresy.[38]

But let us get back to the idea of being 'hopelessly cooped up in an exclusively psychic world.' It turns out that for Jung this contradictory notion goes hand in hand with another idea. Thus,

36. *MDR*, pp 351–52.
37. *CW*, vol. 9, pt. I, p 173.
38. *MM*, pp 175–76.

immediately after telling us that 'the psyche cannot leap beyond itself,' he goes on to say that 'it cannot set up any absolute truths, for its own polarity determines the relativity of its statements.'[39]

But this, too, is an antinomial claim. Obviously! For it is plain that the given assertion has itself been put forward as an absolute truth: if it is true, therefore, it nullifies itself. 'Its initial absurdity,' as Frithjof Schuon remarks with reference to statements of this type, 'lies in the implicit claim to be unique in escaping, as if by enchantment, from a relativity that is declared alone to be possible.'[40]

Apparently it does not bother Jung that he contradicts himself at every turn. Perhaps, once one has swallowed the idea that God Himself is the epitome of contradiction, such conduct may seem positively virtuous.

ONE IS INCLINED to agree with Philip Sherrard and others who suggest that Jung's primary objective was the dethronement of Christianity and its replacement by a new psychological brand of religion. All the signs point in that direction, and even the most bizarre and contradictory aspects of Jung's teaching fall readily into place once they are viewed in the light of this hypothesis.

It is clear, to begin with, why Jung should have elected to clothe his message in a scientific garb. As he tells us himself, while commenting on the didactic ambitions of Sigmund Freud: 'Today the voice of one crying in the wilderness must necessarily strike a scientific tone if the ear of the multitude is to be reached.' It is not surprising, moreover, that the 'scientific tone' should be particularly conspicuous in Jung's earlier writings, produced during a period in which the young psychiatrist was working to establish himself as a reputable thinker. By the time we arrive at his later literary productions, on the other hand, we sense an increasingly mystical and overtly religious cast. 'Yet he waited,' as Philip Rieff observes, 'until he was beyond the reach of skeptical reviewers before he published

39. *MDR*, p350.
40. *Logic and Transcendence* (New York: Harper & Row, 1975), p7.

the secret of his life: this burden of prophecy with which he had been charged from the time of his earliest remembered dream.'[41]

Another fundamental ingredient of Jung's thought, as we have seen, is the antinomial creed of dogmatic relativism. This, too, constitutes a 'tone' to which the ear of the multitude is nowadays attuned. But what role, precisely, does it play in the economy of the Jungian catechesis? 'Why, in effect,' asks Philip Sherrard, 'is he issuing a dogma—one, it is true, designed to undermine the traditional basis of religious dogma, but no less a dogma on that account?' And the answer, as Sherrard observes, is fairly clear:

> Indeed, it is precisely this, that he did wish to undermine the traditional basis of religious dogma, as well as of all theological thought of the traditional kind. . . . So long as the great structure of Christian doctrine and dogma, regarded as sacred and inviolate, stood in the way, his own ideas could make little progress. But if he could show that this structure shared in all the necessary limitations of human thought as he conceived them, and was in fact essentially subjective and relative and psychic, its authority would be shaken.[42]

One might add that the dogma of relativism has also an important role to play in relation to science itself. For it serves quite obviously to neutralize the materialistic and rationalist claims with which modern science has been associated from the start, and which no less than Christianity stand in the way of the new religion. The latter demands that not only the Christian God and all the traditional metaphysical categories, but also the physical universe itself should ultimately be swallowed up by the Unconscious, which has been earmarked to play the part of a pantheistic Godhead in Jung's 'theology'. Thus, when Jung confides to us in his posthumous memoirs—by way of interpreting one of his prophetic dreams— that 'our unconscious existence is the real one and our conscious world is a kind of illusion, an apparent reality constructed for a specific purpose, like a dream which seems a reality as long as we are in

41. *The Triumph of the Therapeutic* (New York: Harper & Row, 1968), p110.
42. *Studies in Comparative Religion* 3 (1968), p35.

it,'[43] we are clearly approaching the bottom line of his teaching: it amounts to a psychologization of the Vedantic position which spuriously reduces the conception of *Brahman* to the collective unconscious.

But let us return to Jung's dialectic. After having deposed, at one stroke, the absolutist claims of both traditional metaphysics and modern science, Jung proceeds to preach his own doctrine, not as a metaphysical dogma, or even as a well-substantiated scientific theory, but in ostentatiously tentative terms. 'It seems hardly necessary to add,' he tells us, 'that I hold the truth of my own views to be equally relative, and regard myself also as the exponent of a certain predisposition.'[44] He has no absolute truths to proclaim, Jung avows repeatedly, and he does not presume to encroach upon theological or metaphysical territory. 'In other words,' says Sherrard, 'his system of thought could claim validity not because it was metaphysical, but precisely because it was not metaphysical.'[45]

Yet, once that claim to validity had been more or less accepted, Jung was quite obviously willing to dispense with these epistemological niceties and get down to the point. In his polemics against the *privatio boni*, for example, he seems to forget all about his relativism: when it comes to the Christian belief that God constitutes the *Summum Bonum*, he perceives—not a relative truth or 'a certain predisposition'—but simply a 'false conclusion'. Nor do we detect so much as a trace of relativism when Jung is expounding his own rather mystical conclusions; when he declares, for instance, with reference to the psyche: 'Not only is it the condition of all metaphysical reality, it is that reality.' Obviously there is nothing here to soften the dogmatic thrust of this pronouncement. These statements are made from on high, and are apparently received as such by the faithful. One senses that it is in the form of a psychological quasi-mysticism that Jung's teaching attains its true end.

Jung comes close to saying as much in his autobiography, a work which, more than any other, provides us with a vivid insight into

43. *MDR*, p324.
44. *MM*, p57.
45. *Studies in Comparative Religion* 3 (1969), p36.

the nature and purpose of his doctrine. To begin with, it paints the intellectual and religious background of this enigmatic man, a legacy which Jung himself deems to be of fundamental importance in the shaping of his life's work. Thus it is by no means irrelevant to recall that eight of Jung's uncles were pastors, and that his father, too, was a pastor, who had partially lost his faith and suffered from attacks of insanity which led to his commitment to an asylum. It is clear, moreover, that the religious issue was from the start the central problem which preoccupied the future psychiatrist during his formative years, and as a matter of fact, Jung refers to religious subjects incessantly while recollecting his childhood experiences. One of these was a dream—or was it a vision?—wherein he beheld God 'seated on His golden throne, high above the world,' from whence forthwith 'an enormous excrement' dropped down upon a cathedral, shattering the roof and breaking the walls asunder. Some eighty years later Jung was still able to recollect vividly the impact of this early revelation, the 'unutterable bliss' which he had felt in its wake, and his youthful conviction that 'I had experienced an illumination.'[46] Some time later, we are told, the young seer interpreted this 'illumination' to mean that 'God Himself had disavowed theology and the Church founded upon it.'[47] This was his first prophetic mandate, Jung suggests, the first time—but not by any means the last—that God spoke to him. Thus favored and enlightened, as he sincerely believed, the boy was apparently able to resolve to his own satisfaction the religious perplexities which he had witnessed in his father. This he did by developing a counter-position to Christianity, a work that was to be the great passion of his life. 'Thus Jung found his way out of the religious impasse which had destroyed his father,' as Rieff notes, 'in an integrative personal symbolism, a meta-religion, revealed originally to himself alone, which he then translated, without disclosing its divine source, into a psychotherapy. . . .'[48]

Yet for all its syncretistic tendencies and Oriental borrowings, it appears that this meta-religion retains a certain affinity with

46. *MDR*, p.40.
47. Ibid., p 93.
48. *The Triumph of the Therapeutic* (cited in n 186 above), p 113.

Christianity: the final product of Jung's thought still reflects its Christian starting point. Only the reflection turns out to be an inverted one: 'He has supplied a parody of Christianity,' writes Rieff, 'stopping short of his own "Christification".'[49] But not for long; for as Rieff astutely observes: 'To avoid martyrdom, Jung delayed announcing his full membership in the confraternity of prophets until after his death, by arranging a posthumous publication of his autobiography, which is at once his religious testament and his science, stated in terms of a personal confession.'[50]

IN THE FINAL ANALYSIS, what Jung has to offer is a religion for atheists and a mysticism for those who love only themselves. On the one hand, he extols what he terms the religious attitude as 'an element in psychic life whose importance can hardly be overrated,' while affirming, at the same time, that 'the psychologist of today ought to realize once and for all that we are no longer dealing with questions of dogma and creed.'[51] In other words, it does not matter whether the objective content of religious belief is true or false: what counts is our subjective religious attitude, and presumably the sense of well-being which this is supposed to engender. It would seem that Jung has discovered the secret of cultivating religious attitudes at will; what in bygone days was acquired at the cost of dogmatic and moral commitments can now be supplied by other means. Yet the new product is not like the old; it is an Ersatz, or as Rieff puts it, 'a religion of a sort—for spiritual dilettants, who collect symbols and meanings as others collect paintings.'[52]

As the matter stands, Jung has ransacked the religions and secret doctrines of the world to provide himself with an impressive pantheon of god-terms. But something is invariably lost in the process. At his touch, the ancient symbols forthwith lose their transcendental significations and acquire a truncated sense: the living God of

49. Ibid., p139.
50. Ibid.
51. MM, p67.
52. The Triumph of the Therapeutic, p139.

Abraham ceases to be the Creator of the universe and becomes simply a father-image, a mere sign standing for an archetype, which is itself no more than a particular content of the collective unconscious. One wonders whether this metamorphosis might not affect the saving efficacy of the religious symbol. But be that as it may, what Jung is passing on to his sophisticated clientele is worlds removed from a religious orientation.

The Jungian archetypes are psychic propensities, as we have seen. Unlike the archetypes of Platonism or of Christianity, they belong to the temporal order and have come into their present state by some historical or evolutionary process. Now if the cosmos is essentially a theophany, as Christian doctrine maintains, then the Jungian archetypes, too, must in a way reflect the eternal 'ideas' which are said to reside in the Logos or Wisdom of God. Only we must not forget that the nature or quality of this reflection depends upon the factor of mental purity: and that is just where the problem lies. None but the 'pure in heart' shall see God. But there is little reason to suppose that the unconscious in its present state, whether private or collective, conforms to exceptionally high standards of purity. Indeed, it may be in worse shape than our conscious mind. Nor is there the slightest reason to believe that the collective unconscious is any better or more spiritual than mankind *per se*, whether we consider this collectivity in its present or in some earlier state of development. Thus, if one assumes the evolutionist claims of progress, it follows that the collective unconscious corresponds to an earlier and consequently lower stage, which the individual of today is called upon to supersede. On the other hand, if religion is right in affirming the fall of man, then it stands to reason that the collective unconscious of a degraded humanity must share in this degradation. In either case, the collective unconscious is certainly not a universal norm or an unfailing source of saving grace as Jung seems to assume. And so far as we can tell, no spiritual tradition on earth has ever claimed as much. Quite to the contrary: we have been severely warned to beware of these murky and ambivalent depths, and of the psychic forces or occult entities which may reside in that nether realm. If there be such a thing as a spiritually legitimate 'harrowing of hell', it is to be approached with fear and trembling, and not without the protection of sacramental grace.

Getting back to the Jungian archetypes, it is unreasonable to maintain that these psychic forms or propensities are quite as immutable as Jung makes them out to be. One must not press the analogy with fossils too far: mind, unlike stone, is an inherently protean element. It is only to be expected, therefore, that the collective unconscious and its so-called archetypes should be changing continually. So far from being perfectly homogeneous in time and in point of ethnic distribution, the collective unconscious is bound to respond somewhat to historical exigencies, and must consequently be subject to local variations. Quite possibly, as Titus Burckhart suggests, it may undergo a certain deterioration within major cultural or ethnic groups, brought on by a collective apostasy from established religious and moral norms. We will quote what Burckhart has to say on this important question:

> In every collectivity that has become unfaithful to its own traditional form, to the sacred framework of its life, there ensues a collapse or a sort of mummification of the symbols it has received, and this process will be reflected in the psychic life of every individual belonging to that collectivity and participating in that infidelity. To every truth there corresponds a formal trace, and every spiritual form projects a psychic shadow; when these shadows are all that remain, they do in fact take on the character of ancestral phantoms that haunt the subconscious. The most pernicious of psychological errors is to reduce the meaning of symbolism to such phantoms.[53]

It is Jung, of course, who has dogmatically reduced the meaning of symbolism to 'such phantoms', as if there were nothing else for religious man to contemplate than the Jungian archetypes. This amounts to a deification of the collective unconscious, and so of man, from whom this unconscious derives and to whom it belongs. In the psychologistic quasi-theology of Jung, the blurred memory of our race has assumed the position of Godhead, and the collective evolving 'self'—whatever that may be—has become the personal God.

53. 'Cosmology and Modern Science', in *The Sword of Gnosis* (Baltimore: Penguin, 1974), pp 174–74.

What makes the Jungian cult of self-worship especially seductive —and perhaps more dangerous to religion than any other ideological system presently in vogue—is its pan-religious and scientific garb, which disarms almost everyone, and has led even a learned Dominican to speak of the Swiss psychiatrist in exuberant tones as 'a priest without a surplice'. In any case, Jung's influence upon Christianity is definitely on the upswing. And as might be expected, it is precisely among the religious intellectuals and spiritual seekers that this influence is most pronounced. Here at last is an anti-creed that could indeed 'deceive even the elect'! Moreover, when it comes to ecclesiastics whose bent may be less mystical, the Jungian blend of religion and psychotherapy is frequently perceived as the means par excellence by which those 'sophisticated accommodations with the negative communities of the therapeutics' may be brought about. And here is something that is fast moving from the planning stage to the level of implementation: it is already happening. In churches all over the land it would appear that Jung has already gained admittance into the sanctuary.

7

'PROGRESS'
IN RETROSPECT

EVERY AGE, EVERY CIVILIZATION, has a spirit of its own. It is this that determines the habitual outlook, the typical way of looking at things, the values, norms and interdictions—in short, the essentials of the culture. It is quite certain, moreover, that most individuals will conform to the prevailing tendencies of the civilization into which they have been born, and this applies also no doubt to the majority of those who consider themselves to be non-conformists. On the other hand, it must also be possible to transcend cultural boundaries: there can really be no such thing as a rigid cultural determinism. But yet this crossing of boundaries turns out to be a rather rare occurrence; it happens much less frequently than we are led to suppose. We must not let ourselves be fooled. It is true, for example, that in modern times there has been an unprecedented interest in the study of history; and yet one finds that it is almost invariably a case of history truncated by the mental horizon of our age and colored by the humanistic sentiments of our civilization. The Zeitgeist is indeed a force to be reckoned with, and it is never easy to swim against the stream.

Yet this is precisely what must be done if we are to gain an unbiased perspective on the modern world. To put it rather bluntly, we need to break out of the intellectual smugness and provincialism of the typically modern man, the individual who has become thoroughly persuaded that our civilization represents the apex of a presumed 'human evolution', and that mankind had been groping in darkness until Newton and his scientific successors arrived upon the scene to bring light into the world. Now this is not to deny that bygone ages have known their share of ignorance and other ills, and that in certain respects the human condition may have been improved. Our point, rather, is that these supposedly positive developments which figure so prominently in the contemporary perception of history represent only a part of the story: the lesser part, in fact. We see the things that we have gained and are blind—almost by definition—to all that has been lost. And what is it that has been lost? Everything, one could say, that transcends the corporeal and psychological planes, the twin realms of a mathematicized objectivity and an illusory subjectivity. In other words, as intellectual heirs to the Cartesian philosophy we have become denizens of an impoverished universe, a world whose stark contours have been traced for us by the renowned French rationalist. At bottom there is physics and there is psychology—answering to the two sides of the great Cartesian divide—and together the two disciplines have in effect swallowed up the entire locus of reality: our reality, that is. Beyond this we see nothing; we cannot—our premises do not permit it.

But what then is out there that could possibly be seen? And by what means? The answer is surprisingly simple: what is to be seen is the God-made world, and this seeing—this prodigy—is to be accomplished through the God-given instruments consisting of the five senses and the mind. In this way we actually come into contact with the real, objective cosmos, which turns out to be a live universe full of color, sound and fragrance, a world in which things speak to us and everything has meaning. But we must learn to listen and to discern. And that is a task which involves the whole man: body, soul, and above all, 'heart'. Everyone has seen a bird or a cloud, but not everyone is wise, not everyone is an artist in the true sense. This is of course what an education worthy of the name should help us to achieve: it should make us wise, it should open the eye of the soul.

One question remains: what is it that Nature has to tell—if only one has 'ears to hear'? Now to begin with it speaks of subtle things, of invisible causes and of cosmic harmonies. There is a science to be learned, a 'natural philosophy' that is not contrived. But that is not all; it is only the merest beginning. For at last—when 'the heart is pure'—we discover that Nature speaks, not of herself, but of her

Maker: 'Heaven and earth are full of Thy glory.' Or in the words of the Apostle, '*the invisible things of Him from the creation of the world have been clearly seen, being understood by the things that are made, even His eternal power and Godhead.*'

But as we are well aware, the very recollection of this exalted knowledge began to wane long ago and by the time of the Renaissance had grown exceedingly dim, except in the case of a few outstanding souls. When it comes to Galileo and Descartes, moreover, it would appear that the light had gone out entirely: their philosophy of Nature leaves little room for doubt on that score. And from here on one encounters a prevailing intellectual milieu that is truly benighted, whatever the history books may say. To be sure, there have been some notable voices crying in the wilderness, and yet it is plain to see that 'Bacon and Newton, sheath'd in dismal steel' have carried the day, and that their 'Reasonings like vast Serpents' have infolded 'the Schools and Universities of Europe,' as Blake laments to his everlasting glory. It was the victory of 'single vision': a kind of knowing which paradoxically hinges upon a scission, a profound alienation between the knower and the known. Now this is the decisive event that has paved the way to modern culture. From that point onwards we find ourselves (intellectually) in a contrived cosmos, a world cut down to size by the profane intelligence—a man-made universe designed to be comprehensible to physicists, and for its very lack of objective meaning, to psychologists as well.

Or this is where we would find ourselves, better said, if the great modern movement had fully succeeded in converting us to its preconceived notions. But that is not really possible; on closer examination we are bound to discover that there is in fact no one on earth who fully believes—with all his heart—what science has to say: such a Weltanschauung can speak only to a part of us, to a single faculty as it were, and so it is in principle unacceptable to the total man. Still, there is no denying that collectively we have become converts to a high degree. And if the vision does not fit the whole man, he can learn to live piecemeal, by compartments so to speak. Having become alienated from Nature—the object of knowledge—he becomes in the end estranged from himself.

We are beginning to see that the cosmological train of thought which started idyllically enough with the garden meditations of Descartes has had cultural reverberations. Roszak is unquestionably right when he insists that 'cosmology implicates values', and that 'there are never two cultures; only one—though that one culture may be schizoid.'[1] He may also be right when he speaks of the outward consequences of this cultural neurosis in the following terms:

> We can now recognize that the fate of the soul is the fate of the social order; that if the spirit within us withers, so too will all the world we build about us. Literally so. What, after all, is the ecological crisis that now captures so much belated attention but the inevitable extroversion of a blighted psyche? Like inside, like outside. In the eleventh hour, the very physical environment suddenly looms up before us as the outward mirror of our inner condition, for many the first discernible symptom of advanced disease within.[2]

FOLLOWING UPON these summary observations, it may be well to reflect on the first major achievement of modern science, which is no doubt the Copernican astronomy. One generally takes it for granted that the displacement of the Ptolemaic by the Copernican world-view amounts to a victory of truth over error, the triumph of science over superstition. There are even those who perceive the Copernican position as a kind of holy doctrine having Giordano Bruno as its martyr and Galileo as its saintly confessor. But strangely enough it is forgotten that twentieth-century physics is in fact neutral on the entire issue. There was first of all the question whether the sun moves while the Earth remains fixed, or whether it is really the Earth that moves, and not the sun. Now what modern physics insists upon—ever since Einstein recognized the full implication of the Michelson-Morley experiment—is that the concepts of rest and motion are purely relative: it all depends on what we take to be our frame of reference. Thus, given two bodies in space, it makes no

1. *Where the Wasteland Ends* (Garden City, NY: Doubleday, 1973), p 200.
2. Ibid., p xvii.

sense whatever to ask which of the two is moving and which is at rest. So much for the first point of contention. The second issue, moreover, related to the position of the two orbs, each side claiming that the body which they took to be at rest occupies the center of space. And here again contemporary physics sees a pseudo-problem arising from fallacious assumptions. The question is in fact senseless on two counts: first, because (as we have seen) one cannot say that a body is at rest in an absolute sense; and secondly, because there is actually no such thing as a center of space. Thus, whether one conceives of cosmic space as unbounded (like the Euclidean plane) or as bounded (like the surface of a sphere), there exists in either case no special point that is marked out from the rest, and so also no point which could be taken as the center of space. But in the absence of a center the Copernican debate loses its meaning; from this perspective the entire controversy appears indeed as the classic example of 'much ado about nothing'.

Yet this way of looking at the matter—which equalizes the two contesting sides—turns out to be no less deceptive than the popular view which bestows the palm of victory on the Copernicans. If the popular verdict is based on little more than prejudice and propaganda, the scientific appraisal for its part rests on the no less gratuitous assumption that cosmology is to be formulated in purely quantitative and 'operationally definable' terms. One tacitly assumes, in other words, that quantity is the only thing that has objective reality, and that the *modus operandi* of empirical science constitute the only valid means for the acquisition of knowledge. Now historically this is just the position to which Western civilization has been brought through a series of intellectual upheavals and reductions in which the Copernican revolution has played a major role. In fact, the new outlook stems directly from the later Copernicans, individuals like Galileo, whose thought was already modern in that regard. One should also remember that these (and not Copernicus) are the men who ran afoul of the ecclesiastical authorities and precipitated the famous debates. It was in the year 1530, let us recall, that Copernicus communicated his ideas to Pope Clement VII and was encouraged by the Pontiff to publish his inquiries; and it was a century later (in the year 1632) that Galileo was summoned before the Inquisition. The point is that there was more to the celebrated controversy than first meets the eye; and while overtly the debate raged over such seemingly harmless issues as whether it is the Earth or the sun that moves, one can see in retrospect that what was actually at stake was nothing less than an entire Weltanschauung.

We tend to forget that the Ptolemaic world-view was incomparably more than simply an astronomical theory in the contemporary sense; we forget that it was a *bona fide* cosmology as distinguished from a mere cosmography of the solar system. Now to appreciate the point of this difference, it must be recalled that the ancient Weltanschauung conceives of the cosmos as an hierarchic order consisting of many 'planes', an order in which the corporeal world—made up of physical bodies, or of 'matter' in the sense of modern physics—occupies precisely the bottom rank. This implies, in particular, that whatever can be investigated by the methods of physics—everything that shows up on its instruments—belongs *ipso facto* to the lowest fringe of the created world. Newton *was* right: we *are* only gathering pebbles by the seashore; for indeed, the physical sciences, by their very nature, are geared to the corporeal order of existence. Now basically this is just the world that is perceptible to our external senses; only we must remember that even this lowest tier of the cosmic hierarchy is incomparably richer than the so-called physical universe—the ideal or imagined cosmos of contemporary science—because, as we have had ample opportunity to see, the corporeal world comprises a good deal more than simply mathematical attributes. Thus, if we wanted to locate the universe of modern physics on the ancient maps, we would have to say that it constitutes an abstract and exceedingly partial view of the outermost fringe, the 'shell' of the cosmos. A *bona fide* cosmology, on the other hand, in the traditional sense, is a doctrine that bears reference—not just to a single plane—but to the cosmos in its entirety.

The question arises, of course, how the Ptolemaic theory, which after all does speak of the sun and its planets, could 'bear reference to the cosmos in its entirety,' seeing that the corporeal order as such constitutes no more than the smallest part of that total cosmos. And the answer is simple enough, at least in principle: the things of Nature point beyond themselves; though they be corporeal, they speak of incorporeal realms—they are symbols. In fact, there is an analogic correspondence between the different planes: 'as above, so below' says the Hermetic axiom. We must not forget that despite its hierarchic structure the cosmos constitutes an organic unity, much like the organic unity of mind, soul and body which we can glimpse within ourselves. Does not the face mirror the emotions or thoughts, and even the very spirit of the man? We have become oblivious of the fact that the cosmos, too, is an 'animal', as the ancient philosophers had observed.

This, then—the miracle of cosmic symbolism—is what stands behind the Ptolemaic world-view and elevates it from a somewhat crude cosmography to a full-fledged cosmology. There was a time, moreover, when men could read the symbol, when they sensed that the solid Earth as such represents the corporeal realm, which stands at the very bottom of the cosmic scale; and that beyond this Earth there are spheres upon spheres, each larger and higher than the one before, until one arrives at last at the Empyrean, the ultimate limit or bound of the created world. They sensed too that there is an axis extending from Heaven to Earth, by which all these spheres are held together as it were, and around which they revolve; and they realized intuitively that the relation of containment is expressive of preeminence: it is the higher, the more excellent, that contains the lower, even as the cause contains the effect or the whole contains the part.

Let us add that in attempting to appraise these ancient beliefs we must not be put off by the fact that their erstwhile proponents— men who supposedly had some intuitive apprehension of higher realms—were evidently ignorant of things that are nowadays known to every schoolboy. We need not be unduly astonished, for example, that Ptolemy took our planet to be fixed in space because 'if there were motion, it would be proportional to the great mass of the Earth and would leave behind animals and objects thrown into the air.'[3] Childish, yes; but we should remember that the Book of Nature can be read in various ways and on different levels, and that

3. Quoted by E.A. Burtt, *The Metaphysical Foundations of Modern Physical Science* (New York: Macmillan, 1951), p35.

no one knows it all. To be sure, 'There are more things in heaven and earth, Horatio, than are dreamt of in your philosophy.'

Getting back to the Copernican debate, it has now become apparent that the change from a geocentric to a heliocentric astronomy is not after all such a small or harmless step as one might have imagined. The fact is that for all but a discerning few it has undermined and discredited a cosmic symbolism which had nurtured mankind throughout the ages. Gone was the visible exemplification of higher realms and the vivid sense of verticality which spoke of transcendence and of the spiritual quest. Gone was the world that had inspired Dante to compose his masterpiece. With the demise of the Ptolemaic world-view the universe was in effect reduced to a single horizontal cross-section—the lowest, no less. It has become for us this narrow world, which remains so for all the myriad galaxies with which we are currently being regaled. Nature has become 'a dull affair', as Whitehead says, 'merely the hurrying of material, endlessly, meaninglessly.'

One might object to this assessment of what was actually at stake in the Copernican issue on the grounds that a heliocentric astronomy too admits of a symbolic interpretation, since it identifies the sun—a natural symbol of the Logos—as the center of the cosmos. But yet the fact remains that its rediscovery by Copernicus has not been propitious to a spiritual vision of the world; 'rather was it comparable to the dangerous popularization of an esoteric truth', as Titus Burckhardt observes.[4] One must remember that our normal experience of the cosmos is obviously geocentric, a fact which in itself implies that the Ptolemaic symbolism is apt to be far more accessible. Moreover, the Copernican victory came at a time when the religious and metaphysical traditions of Christianity had already fallen into a state of partial decay, so that there was no longer any viable framework within which the symbolic content of heliocentrism could have been brought to light. As Hossein Nasr has pointed out, 'the Copernican revolution brought about all the spiritual and religious upheavals that its opponents had forecasted

---

4. 'Cosmology and Modern Science', in *The Sword of Gnosis* (Baltimore: Penguin, 1974), p127.

would happen precisely because it came at a time when philosophical doubt reigned everywhere....'[5] It was a time when European man was no longer especially attuned to the reading of transcendental symbols and had already to a large extent lost contact with the higher dimensions of existence. And this is what lends a certain air of unreality to the Copernican dispute, and what from the start assured the eventual triumph of the new orientation. By now the wisdom of bygone ages—like every truth that is no longer understood—had become a superstition, to be cast aside and replaced by new insights, new discoveries.

WITH THE DISAPPEARANCE of the Ptolemaic world-view Western man lost his sense of verticality, his sense of transcendence. Or rather these finer perceptions had now become confined to the purely religious sphere, which thus became isolated and estranged from the rest of the culture. So far as cosmology—Weltanschauung in the literal sense—was concerned, European civilization became de-Christianized.

At the same time a radical change in man's perception of himself was taking place. We need to recall in this connection that according to ancient belief there is a symbolic correspondence between the cosmos in its entirety and man, the theomorphic creature who recapitulates the macrocosm within himself. Thus man is indeed a 'microcosm', a universe in miniature; and that is the reason why, symbolically speaking, man is situated at the very center of the cosmos. In him all radii converge; or better said, from him they radiate outwards in every direction to the extremities of cosmic space—a mystical fact which we find graphically depicted in many an ancient diagram. No doubt the reason for this centrality is that man, having been made 'in the image of God', carries within himself the center from which all things have sprung. And that too is why he can understand the world, and why in fact the cosmos is intelligible to the human intellect. He is able to know the universe because in a way it pre-exists in him.

5. *Man and Nature* (London: Allen & Unwin, 1976), p66.

But of course all this means absolutely nothing from the modern point of view. To be sure, once the cosmos has been reduced to the corporeal plane, and that in turn has been cut down to its purely quantitative parameters, there is little left of the aforementioned analogy. Admittedly our physical anatomy does *not* resemble the solar system or a spiral nebula. It is first and foremost in the qualitative aspects of creation, as revealed to us through the God-given instruments of perception, that cosmic symbolism comes into play. We need not be surprised, therefore, that a science which peers upon Nature through lifeless instruments fashioned by technology should have little to say on that score.

In any case, along with the Ptolemaic theory the ancient anthropology fell likewise into oblivion. Man ceased in effect to be a microcosm, a theomorphic being standing at the center of the universe, and became instead a purely contingent creature, to be accounted for by some sequence of terrestrial accidents. Like the cosmos he was flattened out, shorn of the higher dimensions of his being. Only in his case it happens that 'mind' refuses to be altogether exorcised. It remains behind as an incomprehensible concomitant of brain-function, a kind of ghost in the machine, a thing that causes untold embarrassment to the philosophers. The fact is that man does not fit into the confines of the physical universe. There is another side to his nature—be it ever so subjective!—which cannot be described or accounted for in physical terms. And so, in keeping with the new outlook, man finds himself a stranger in a bleak and inhospitable universe; he has become a precarious anomaly—one could almost say, a freak. There is something pathetic in the spectacle of this 'precocious simian'; and behind all the noise and bluster one senses an incredible loneliness and a pervading *Angst.* Our harmony and kinship with Nature has been compromised, the inner bond broken; our entire culture has become dissonant. Moreover, despite our boast of knowledge, Nature has become unintelligible to us, a closed book; and even the act of sense perception—the very act upon which all our knowledge is supposed to be based—has become incomprehensible.

What then are we to say concerning the stupendous knowledge of science? It is evidently a knowledge that has been filtered through external instruments and that partakes of the artificiality of these man-made devices. Strictly speaking, what we know is not Nature but certain methodically monitored effects of Nature upon that mysterious entity termed 'the scientific observer'. It is thus a postivistic knowledge geared to the prediction and control of phenomena, and ultimately—as we know—to the exploitation of natural resources and the practice of terrestrial rapine. All euphemisms aside, science—like most else that modern man busies himself with—is well on the way to becoming simply an instance of 'technique' in the sense of the sociologist Jacques Ellul.

Meanwhile all the ideal aspects of human culture, including all values and norms, have become relegated to the subjective sphere, and truth itself has become in effect subsumed under the category of utility. Transcendence and symbolism out of the way, there remains only the useful and the useless, the pleasurable and the disagreeable. There are no more absolutes and no more certainties; only a positivistic knowledge and feelings, a veritable glut of feelings. All that pertains to the higher side of life—to art, to morality or to religion—is now held to be subjective, relative, contingent—in a word, 'psychological'. One is no longer capable of understanding that values and norms could have a basis in truth. How could this be in a world of 'hurrying material'? And so man has become the great sophist: he has set himself up as 'the measure of all things'. Having but recently learned to walk on his hind legs (as he staunchly believes), he now fancies himself a god! 'Once Heaven was closed,' writes Schuon, 'and man was in effect installed in God's place, the objective measurements of things were, virtually or actually, lost. They were replaced by subjective measurements, purely human and conjectural pseudo-values.'[6]

Thus, too, all the elements of culture, having once been subjectivized, have become fair game to the agents of change. Nothing is sacrosanct any more, and at last everyone is at liberty to do as he will. Or so it may seem; for in reality the manipulation of culture has become a serious enterprise, a business to be attended to by governments and other interest groups.

6. *Light on the Ancient Worlds* (London: Perennial Books, 1965), p30.

We find thus that cosmology does indeed 'implicate values'; one could even say that eventually it turns into politics. So too a pseudo-cosmology necessarily implicates false values, and a politics destructive of good. It is by no means a harmless thing to be cut off from the higher spheres or from the mandates of God. Our civilization has forgotten what man is and what human life is for; as Nasr notes, 'there has never been as little knowledge of man, of the *anthropos*.'[7] To which one might add that apparently no previous culture has managed to violate so many natural and God-given norms to any comparable extent.

SOME REFLECTIONS on the subject of art may not be inappropriate at this point. The first thing to be noted is that the very conception of art has changed: the word has actually acquired a new meaning. Thus art has become 'fine art', something to be enjoyed in leisure moments and generally by the well-to-do. It has become a luxury, almost a kind of toy. In ancient times, on the other hand, 'art' meant simply the skill or wisdom for making things, and the things made by art were then called 'artefacts'. Strictly speaking everything that answered a legitimate need and that had to be produced by human industry was an artefact. Thus an agricultural implement or a sword was an artefact, a piece of furniture or a house was an artefact, and so too was a cathedral or an icon or an ode. The artefact, moreover, was there for the whole man, the trichotomous being made up of body, soul, and spirit; and so even the humblest tool or utensil had to possess more than simply 'utility', in the contemporary sense. That 'more', of course, derives from symbolism, from the language of forms. It is the reason why a water-pot can be a thing of immense beauty and meaning. Not that this beauty had to be somehow superimposed upon the object, like an ornament. It was there as a natural concomitant of utility, of the 'correctness', one could say, of the work. And that is the reason why in ancient times there was an intimate link between art and science, and why Jean Mignot (the

---

7. 'Contemporary Man, between the Rim and the Axis', *Studies in Comparative Religion*, 7 (1973), p116.

builder of the cathedral at Milan) could say that 'art without science is nothing' (*ars sine scientia nihil*). In a word, both beauty and utility were conceived to spring from truth.

It was understood, moreover, that authentic art can never be profane. For let us remember that according to Christian teaching the eternal Word or Wisdom of God is indeed the supreme Artist: '*All things were made by Him, and without Him was not anything made.*' Now it follows from the profound sense of this text that whatever is truly made, or made rightly, is made by Him. And this implies that the human artist—every authentic artist—must participate to some degree in the eternal Wisdom. 'So, too, the soul can perform no living works,' writes St Bonaventure, 'unless it receive from the sun, that is, from Christ, the aid of His gratuitous light.'[8] Man, therefore, the human artist, is but an agent; to achieve perfection in his art he must make himself an instrument in the hands of God. And so the production of the artefact is to be ascribed to the divine Artificer in proportion as it is beneficent and well made; for indeed '*every good gift and every perfect gift is from above, and cometh down from the Father of lights*' (James 1:17).

To some extent this constitutes a universal doctrine that has guided and enlightened the arts of mankind right up to the advent of the modern age. Thus even in the so-called primitive societies all art, all 'making', was a matter of 'doing as the gods did in the beginning.' And that 'beginning', moreover, is to be understood in a mythical, that is to say, in a metaphysical sense. Basically it is the ever-present 'now', that elusive point of contact between time and eternity which is also the center of the universe, the 'pivot around which the primordial wheel revolves.' As Mircea Eliade has amply demonstrated, the traditional cultures have been cognizant of that universal center and have sought by ritual or other symbolic means to effect a return to that point of origin, that 'beginning'. That is where man was able to renew himself; from thence he derived strength and wisdom. And from thence too, needless to say, he derived his artistic inspiration. Thus, strange as it may sound, the traditional artist works not so much in time as in eternity. His art

---

8. *De Reductione Artium ad Theologian*, 21.

partakes somehow of the instantaneous 'now'; and this explains its freshness, the conspicuous unity and animation of its productions. No matter how long it may take to fashion the external artefact, the work has been consummated internally in a trice, at a single stroke.

The Scholastics were no doubt heirs to this immemorial conception of art. It is evidently what St Thomas has in mind when he defines art as 'the imitation of Nature in her manner of operation'[9]; for we must understand that here the term 'Nature' is employed not in the current sense—not in the sense of *natura naturata*, a nature that has been made—but in the sense of *natura naturans*, the creative agent which is none other than God. The human artist thus imitates the divine Artificer; for in imitation of the Holy Trinity he works 'through a word conceived in his intellect' (*per verbum in intellectu conceptum*),[10] which is to say, through a word or 'concept' which mirrors the eternal Word. Man too 'begets a word' in his intellect; and this constitutes the *actus Primus* of artistic creation.

It follows from these considerations that there is a profound spiritual significance both in the enjoyment and in the practice of authentic art. On the one hand, a *bona fide* artefact will possess a certain charisma, a beauty and significance which no profane or merely human art could effect—not to speak of mechanized production. Such an artefact will exert an invisible influence upon the user; it will benefit the patron in unsuspected ways. But what is still more important, the exercise of his art will bring not only material remuneration but also spiritual benefit to the artist. 'Manufacture, the practice of an art,' writes Coomaraswamy, 'is thus not only the production of utilities but in the highest possible sense the education of men.'[11] It is a spiritual way, a means to perfection. And one could even say that the practice of an art should be a normal and integral part of the Christian life: everyone should be an artist of some kind, each in accordance with his vocation. As William Blake has expressed it, 'The Whole Business of Man Is the Arts.... The unproductive Man is not a Christian.'

9. *Summa Theologiae*, I, 117, 1.
10. Ibid., I, 45, 6.
11. *Christian and Oriental Philosophy of Art* (New York: Dover, 1956), p 27.

One also knows, however, that as Blake was writing these lines the Industrial Revolution was gathering momentum and the Arts were on their way out. The machine age was upon us, and that kind of manufacture which had been so much more than the mere 'production of utilities' was fast being replaced by the assembly line. We know that efficiency has been increased a hundredfold and that the 'standard of living' has never been so high. And we know too that the promised utopia has not arrived, and that unforeseen difficulties are cropping up at an accelerating pace. What we generally don't know, however, is that our civilization has become culturally impoverished to an alarming degree. We are beginning to become cognizant of the ecological crisis and shudder at the reports of acid rain, but still fail to behold the spiritual wasteland that has been forming around us for centuries. We speak of 'the dignity of labor' and forget that there was a time when manufacture was more than a tedium, a meaningless drudgery which men endure only for the sake of pecuniary reward. We speak of 'the abundant life' and forget that happiness is not simply play, entertainment or 'getting away from it all', but the spontaneous concomitant of a life well lived. We forget that pleasure does not come in pills or via an electronic tube but through what the Scholastics termed 'proper operation', the very thing that authentic art is about. In short, what we have totally forgotten is that 'The Whole Business of Man Is the Arts.'

Besides industry, of course, our culture comprises also 'the fine arts', which are there presumably to supply 'the higher things of life'. Now whatever else might be said in behalf of these productions, it is clear that for the most part they are bereft of any metaphysical content. Our art ceased long ago to be a 'rhetoric' and became an 'aesthetic', as Coomaraswamy has pointed out; which is to say that it is no longer intended to enlighten but only to please. It is not the function of our fine arts 'to make the primordial truth intelligible, to make the unheard audible, to enunciate the primordial word, to represent the archetype,' which from a traditional point of view is indeed 'the task of art, or it is not art,' as Walter Andrae observes.[12] And however sublime this 'fine art' may be, it does not in fact bear

---

12. Quoted by A.K. Coomaraswamy, op. cit., p55.

reference to 'the invisible things of Him' because the artist who made it was simply a man—a genius, perhaps, but a man nonetheless. Unlike ancient art it does not derive 'from above', nor does it refer to spiritual realities, or to God 'whom we never mention in polite society.' As a matter of fact, in keeping with the overall subjectivist trend of modern culture, art has become more and more a matter of 'self-expression', right up to the point where the contingent, the trivial and the base have all but monopolized the scene. A stage has been reached where much of art is plainly subversive—one needs but to recall those bizarre paintings of patently Freudian inspiration which could very well have originated within the walls of a lunatic asylum! The history of modern art teaches us that the merely human, cut off from spiritual tradition and the touch of transcendence, is unstable; it degenerates before long into the infra-human and the absurd.

THERE IS AN INTIMATE CONNECTION between the machine metaphor as a cosmological conception and the creation of a technological society. Let us not forget that a machine has no other *raison d'être* than to be used. When Nature, therefore, is viewed as being nothing more than a machine, it will as a matter of course come to be regarded simply as a potential object of exploitation, a thing to be used in all possible ways for the profit of men. The two attitudes, moreover, go hand in hand; for as Roszak points out, 'only those who experience the world as dead, stupid, or alien and therefore without a claim to reverence, could ever turn upon their environment...with the cool and meticulously calculated rapacity of industrial civilization.'[13] It is therefore not surprising that no sooner had the postulate of cosmic mechanism gained official recognition than men began on an unprecedented scale to build their own machines with which to harness the forces of Nature; in the wake of the Enlightenment came the Industrial Revolution.

But the story does not end there; for it was inevitable within the perspective of the new cosmology that man, too, should come to be

---

13. *Where the Wasteland Ends* (cited in n 199 above), pp 154–55.

viewed as a kind of machine. What else could he be in a Newtonian universe? And if man is a machine, society too is a machine and human behavior is deterministic: Newton, Lamettrie, Hobbes, and Pavlov clearly lie on a single trajectory. And these recognitions—or better said, these new premises—open up incalculable possibilities! Whether we realize it or not, the cold and rigorous dialectic of science in its concrete actuality leads step by step to the formation of a technological society in the full frightening sense of that term.

Let us consider the matter a little more carefully. To understand the scientific process we need to recall an essential idea which goes back not so much to Newton as to Descartes, and is especially associated with the name of Francis Bacon (the first of the two 'archvillains' in Blake's vision of Victorious Science). Now Bacon's contribution resides in his perception of a universal and all-encompassing method for the systematic acquisition of knowledge. In the first place, this process is envisaged as collective and cumulative; it is an enterprise that keeps on gathering momentum. Thus 'the business' of knowing should not be left in the hands of the individual but is to be carried out by teams of experts, as we would say; and significantly enough (this is its second notable characteristic), it is to be done 'as if by machinery'. Here it is again: the all-conquering omnivorous machine metaphor! But this time in an entirely new key: as a methodological principle. With telling effect Bacon goes on to observe how very small would be the accomplishments of 'mechanical men' if they worked only with their bare hands, unaided by tools and instruments contrived through human ingenuity. In like manner very little can be accomplished when men seek to acquire knowledge through 'the naked forces of understanding.' In the mental domain, too, we need a tool, an instrument of thought; and that is just what his '*novum organum*'—Bacon's famed method of science—is intended to supply. 'A new machine for the mind', he calls it. And like every machine it is there to be used for profit; truth and utility, he assures us, 'are here one and the same thing.'

One can say in retrospect that whereas Bacon's specific recipes for scientific discovery have proved to be relatively useless (as many have pointed out), his dream of a systematic and collective science in which 'human knowledge and human power meet in one' has no doubt been realized beyond his wildest expectations. What has triumphed is not so much any specific 'machine for the mind' but the idea of method or technique as something formal and impersonal that interposes itself between the knower and the known. And whereas on the one hand this artificial intermediary has isolated the knower—impeded his direct access to reality—it has also made possible the development of a formal and depersonalized knowledge, based upon the systematic labors of countless investigators. First came the development of classical physics and what might be termed 'hard' technology. Later the modern biological sciences began to emerge, and later still the so-called behavioral and social sciences. Meanwhile the process of scientization began to extend itself beyond the boundaries of every formally recognized science and proceeded to exert a dominant influence in other domains. 'Scientific knowledge becomes, within the artificial environment, the orthodox mode of knowing,' writes Roszak; 'all else defers to it. Soon enough the style of mind that began with the natural scientist is taken up by imitators throughout the culture.'[14] And as the matter stands, this 'style of mind' is to be encountered everywhere; it has entered into cloisters and convents. It has become a mark of enlightenment, the respected thing; 'all else defers to it.' As Bacon had shrewdly seen, there are in principle no limits to the scientization of culture: given free reign, the process is bound to insinuate itself into virtually every sphere of human thought and every activity.

It is obvious to all that our outer life-styles are being drastically altered as a direct consequence of the scientific advance. What we generally fail to realize, on the other hand, is that the impact of this same development on our inner lives—yes, on the condition of our soul—is no less pronounced. To begin with, the mechanization of our work-environment, the phenomenon of urban sprawl, the rising congestion and perpetual noise, the proliferation of concrete, steel and plastic, the loss of contact with Nature and with natural things, the invasion of our homes by the mass media—all this in itself is bound to have its effect on our mental and emotional condition. Add to this the uprooting of people from their ancestral environment, an

14. Ibid., p31.

unprecedented mobility which shuffles populations like a deck of cards! Add also the other innumerable mechanisms within the technological society which tend to break down every natural division and all cultural ties. Let us add up (if we are able!) all the factors which homogenize and level out. For it must not be forgotten that people too have to be standardized, like interchangeable parts of a machine, so that the wheels of the mechanized civilization may run smoothly and efficiently.

It is to be noted, moreover, that in the course of the present century this leveling, which began with the Industrial Revolution, has entered upon a new phase due to the rise of the behavioral and social sciences. Now from a purely academic point of view it may well appear that these disciplines are of little consequence; for apart from the factual information which they have accumulated (much of it in the form of statistical data) it would seem that one can hardly speak of 'science' at all. The trappings of science (fancy terms and reams of computer print-out) are there no doubt, but very little of its substance—so long, at least, as one insists that the objective verification of hypotheses, without obfuscation and fudging, constitutes a *sine qua non* of the scientific process. And this deficiency is occasionally admitted even by members of the profession. There is the case of Stanislav Andreski, for example, who has offered[15] insightful observations on such subjects as 'The Smoke Screen of Jargon', 'Quantification as Camouflage', 'Ideology Underneath Terminology', and most important of all, 'Techno-Totemism and Creeping Crypto-Totalitarianism'. There it is! This is just the point: if we take a closer look at these seeming pseudo-sciences we find that they too fit perfectly into the integral framework of the technological society. Here too one encounters a kind of 'knowledge' which begets power. As we have already seen in the case of Freudian and Jungian psychology, a pseudo-science may not be without its 'utility', its technical efficacy. And if Voltaire could say that even lying becomes 'virtuous' when it is practiced for the right end, then why (in a pragmatic civilization) should not these human techniques be deemed a science and their dogmas 'truth'?

15. *Social Sciences as Sorcery* (London: Deutsch, 1972).

Be that as it may, the fact remains that our century has witnessed a dramatic increase in the utilization on the part of governments, industries and other powerful interest groups of methods based upon the so-called behavioral and social sciences. A well-known story about Pavlov may be recalled in this connection: it is reported that shortly after the Bolshevik Revolution the famed scientist was virtually imprisoned in the Kremlin and ordered to write a book describing in detail how behavioral methods based upon his theory of conditioned reflexes may be applied to the indoctrination and control of human beings. Whether it be true that Lenin, upon reading the book, exclaimed to Pavlov 'you have saved the Revolution!' —one does know with certainty that Pavlovian methods have been used extensively in the Soviet Union, and that similar techniques have also been developed and applied in the Western democracies.[16]

However, this does not preclude the fact that the vast majority of people, be it in Russia or in the United States, are almost entirely unaware of this process and could not even imagine the extent to which it has already influenced their own beliefs and psychic make-up. As Jacques Ellul has pointed out with reference to propaganda as a specific area of human technique:

> Propaganda must become as natural as air or food. It must proceed by psychological inhibition and the least possible shock. The individual is then able to declare in all honesty that no such thing as propaganda exists. In fact, however, he has been so absorbed by it that he is literally no longer able to see the truth. The natures of man and propaganda have become so inextricably mixed that everything depends not on choice or on free will, but on reflex and myth. The prolonged and hypnotic repetition of the same complex of ideas, the same images, and the same rumors condition man for the assimilation of his nature to propaganda.[17]

Much the same could be affirmed, moreover, with regard to many other areas of human technique which are not simply 'propaganda'

---

16. See William Sargant, *Battle for the Mind* (Westwood, CT: Greenwood, 1957).
17. *The Technological Society* (New York: Alfred Knopf, 1965), p366.

in the strict sense. Thus it is only to be expected that in our kind of civilization almost every organized 'encounter'—from kindergarten to post-graduate seminars—will entail an element of concealed indoctrination. As Ellul has shown, virtually all education—on both sides of the Iron Curtain—involves mechanisms of conditioning and control designed to fit the individual into the projects of the society.[18] Even our leisure is 'literally stuffed with technical mechanisms of compensation and integration' which, though different from those of the work environment, are 'as invasive and exacting, and leave man no more free than labor itself.'[19] Within the last decade even religious and priestly retreats have become fair game to the scientific methods of 'sensitivity training'! It is the greatest mistake to think that the technological society can be 'culturally neutral', or that the celebrated 'pluralism' about which one hears so much in Western countries can be anything more than a passing phase or an outright fake. 'Cosmology implicates values'—to say it once more—and without any doubt the manipulation of man, the most vital 'resource' of all, constitutes the ultimate technology.

WHILE IT IS SOCIOLOGICALLY CERTAIN that science begets technology, it also cannot be denied that in its purest form science is simply the pursuit of knowledge for its own sake. Like philosophy, it begins in wonder, or in a certain curiosity about Nature; and especially when it comes to the great scientists—an Einstein or a Schrödinger—one finds that the driving force behind their scientific inquiries is indeed worlds removed from any thought of application. One needs but to recall with how much diffidence and anguish Einstein offered his fateful formula to the service of the Free World when the hard exigencies of the time seemed to demand this step. It is one of the great ironies of fate that the most terrible instruments of destruction have been pioneered by men who above all others loved peace, and that the most powerful means of enslavement owe their existence to some of the greatest champions of human liberty.

18. Ibid., p347.
19. Ibid., p401.

But let us pause to reflect a little on the idea of 'knowledge for its own sake'; our sentiments notwithstanding, might there not be an intrinsic connection between this noble quest and such bitter fruit? Preposterous, the humanist will say; and admittedly it has become an almost universally accepted premise that the unbridled pursuit of knowledge constitutes one of the most beneficial and praiseworthy of human occupations. No one seems to question that 'research' of just about any description is a wonderful thing which in some mysterious way is bound to enhance 'the dignity of man' or 'the quality of life'. Not infrequently one finds individuals of even the most prosaic type waxing eloquent in praise of those who are said to have 'pushed back the frontiers of the unknown.' Our libraries are already filled to bursting with the products of this great passion, and yet the cry is always for more. And even when it is recognized that the fruits of this knowledge—the consequences of its applications—have proved to be equivocal or to threaten the very survival of man—even then it is thought that science as such is in no wise at fault. The blame must always be placed at the door of the avaricious entrepreneur or the unscrupulous politician, or it must lie with the short-sighted members of Congress who are held responsible for the under-funding of research. For indeed all ills resulting from 'research and development' are thought to be curable, homeopathic style, with yet another dose of R & D; no one seems prepared to weigh the possibility that the malaise may actually be due, not to an insufficiency, but to an excess of this factor.

Come what may, pure science—science with a capital S—can do no wrong. It is astounding that in an age of unprecedented skepticism, when immemorial beliefs are being tossed aside like worn toys or blithely held up to public ridicule, one should encounter this virtually limitless faith in the unfailing beneficence of scientific research.

What lies behind this passion for more and more science, more and more technology—this mania, one is tempted to say, which has taken hold of our civilization? Is it indoctrination? Yes, no doubt; but then, who first indoctrinated the educators and the technocrats? It is not really quite so simple. Nor can one expect to understand the phenomenon in depth from the typical perspectives of humanist thought. Has not humanism been closely allied with the scientific mentality from the start? Is not the one as well as the other a characteristic manifestation of the contemporary Zeitgeist? Do they not share a common anti-traditional thrust? Were not both equally implicated, for example, in the French Revolution, when 'the Goddess of Reason' was installed on the high altar of Notre Dame? And have not the two—despite the interlude of Romanticism—stood together on almost every issue? It would appear, then, that there can be no searching critique of science which is not also at the same time a critique of humanism. To go beyond superficial appearances and banalities we must be prepared to step out of the charmed circle of contemporary presuppositions and avail ourselves of the only viable alternative to modern thought: and that is *traditional* thought.

What, then, does traditional teaching have to say on the subject of science? We propose to look at the matter from a specifically Christian point of vantage; and even at the risk of speaking what can only be 'foolishness to the Greeks', we shall attempt to place ourselves in an authentically Biblical perspective. This means in particular that we need to reflect anew on the familiar account in Genesis concerning the 'forbidden fruit' and the fall of Adam, his expulsion from 'the garden of paradise'. Now in the first place we must go beyond the customary explanation of this event, which is based upon an essentially moral as opposed to a metaphysical point of view. It is all well and good to attribute Adam's fall to 'the sin of disobedience', and this no doubt expresses a profound and vital truth. But we must also realize that this line of interpretation, valid though it be, cannot possibly cover the entire ground. For one thing it leaves open the question as to why Adam had been commanded to abstain from this particular fruit in preference to all others, and why the tree which brought forth this forbidden harvest is referred to as 'the tree of the knowledge of good and evil'. It is reasonable to suppose, moreover, that 'the apple of knowledge' was indeed fatal not simply because it was forbidden, but that it was forbidden precisely because it would prove fatal to man. Furthermore, we must not think that the 'good' which was to be known through the eating of this fruit is that true or absolute good which religion always associates with the knowledge of God; and neither must we assume that the 'evil' which comes to be revealed through the same act is something objectively real, something which has been created by God. For indeed the opening chapter of Genesis has already informed us many times over that God had surveyed the entire creation and found it to be 'good'. The knowledge, therefore, that is symbolized by the forbidden fruit is a partial and fragmentary knowledge, a knowledge which fails to grasp the absolute dependence of all things upon their Creator. It is a reduced knowledge which perceives the world not as a theophany but as a sequence of contingencies: not *sub specie aeternitatis* but under the aspect of temporality. And it is only in this fragmented world wherein all things are in a state of perpetual flux that evil and death enter upon the scene. They enter thus, on the one hand, as the inescapable concomitant of a fragmentary knowledge, a knowledge of things as divorced from God; and at the same time they enter as the dire consequences of 'disobedience'—the misuse of man's God-given freedom—and so as 'the wages of sin'.

Thus Adam fell. 'The link with the divine Source was broken and became invisible,' writes Schuon; 'the world became suddenly external to Adam, things became opaque and heavy, they became like unintelligible and hostile fragments.'[20] In other words, the world as we know it came into existence: history began. But that is not the whole story. The Biblical narrative has in fact an extreme relevance to what is happening here and now; for as Schuon points out, 'this drama is always repeating itself anew, in collective history as in the life of individuals.'[21] The fall of Adam, then, is not only a primordial act which antedates history as such, but it is also something which comes to pass again and again in the course of human events. It is re-enacted on a smaller or larger scale wherever men opt for what is contingent and ephemeral in place of the eternal truth.

It appears that a 'fall' of major proportions has in fact taken place roughly between the fourteenth and eighteenth centuries. Even the most casual reading of European history reveals the contours of a gigantic transformation: the old order has crumbled and a new

---

20. *Light on the Ancient Worlds*, p44.
21. Ibid.

world has come to birth. To be sure, this is the cultural metamorphosis which we normally behold under the colors of Evolution and Progress; what we do not perceive, on the other hand, is that we have forfeited our sense of transcendence in the bargain. In other words, we have become sophisticated, skeptical and profane. Much as we might wish to be enlightened, the wisdom of the ages has become for us a superstition, a mere vestige of a supposedly primitive past; or at best it is seen as literature and poetry in the exclusively horizontal sense which we currently attach to these terms. Like it or not, we find ourselves in a desacralized and flattened-out cosmos, a meaningless universe which caters mainly to our animal needs and to our scientific curiosity.

Admittedly there are compensations. Energy has been diverted, so to speak, from higher to lower planes, and this accounts undoubtedly for the incredible vigor with which the modernization of our world has been pressed forward and everything on earth is being visibly transformed. At last man is free to devote himself entirely to the mundane and to the ephemeral portion of himself. And this he does, not only with Herculean effort, but with a kind of religiosity. It is one of the salient features of our time that ephemeral goals and secular pursuits—down to the most trivial and inglorious—have become invested with a sacredness, one could almost say, which in bygone ages had been reserved for the worship of God. But why? What is it all about? 'Equipped as he is by his very nature for worship,' writes Martin Lings,

> man cannot not worship; and if his outlook is cut off from the spiritual plane, he will find a 'god' to worship on some lower level, thus endowing something relative with what belongs only to the Absolute. Hence the existence today of so many 'words to conjure with' like 'freedom', 'equality', 'literacy", 'science', 'civilization', words at the utterance of which multitudes of souls fall prostrate in sub-mental adoration.[22]

Everything depends on how we perceive the world, on the quality, one might say, of our knowledge. Is our vision of the universe

22. *Ancient Beliefs and Modern Superstitions* (London: Perennial, 1965), p.45.

centripetal? Is it oriented towards the spiritual center? Is it informed by a sense of verticality, by an intuition of higher spheres? Or is it, on the contrary, horizontal and centrifugal, a knowledge that faces away from the origin, away from the Source? Now that is the kind of knowing which perpetuates the Fall. Always mingled with delusion, it is a profane wisdom that scatters and leads astray. Moreover, it is something to which we have no right by virtue of what we are; like unassimilable food, its very truth becomes eventually a poison to us. Such a knowledge never enlightens us but only blinds our soul; it shuts the gates of Heaven and opens instead the way to the riches of this earth, along with the untold miseries thereof. The terrible fact is that a Promethean science, a science that would make man the measure and master of all things ('ye shall be as gods'), becomes in the end a curse ('cursed is the ground for thy sake, and in sorrow shalt thou eat of it').

INDEX OF NAMES

CPSIA information can be obtained
at www.ICGtesting.com
Printed in the USA
LVHW092310180819
628100LV00004B/11/P

9 781597 310840